Supercomputers and Their Performance in Computational Fluid Dynamics

Edited by Kozo Fujii

Notes on Numerical Fluid Mechanics (NNFM) Volume 37

Volume 8 Vectorization of Computer Programs with Applications to Computational Fluid Dynamics (W. Gentsch)

Volume 12 The Efficient Use of Vector Computers will Emphasis on Computational Fluid Dynamics (W. Schönauer / W. Gentzsch, Eds.)

Volume 14 Finite Approximations in Fluid Mechanics (E. H. Hirschel, Ed.)

Volume 17 Research in Numerical Fluid Dynamics (P. Wesseling, Ed.)

Volume 18 Numerical Simulation of Compressible Navier-Stokes Flows (M. W. Bristeau / R. Glowinski / J. Periaux / H. Viviand, Eds.)

Volume 20 Proceedings of the Seventh GAMM-Conference on Numerical Methods in Fluid Mechanics (M. Deville, Ed.)

Volume 22 Numerical Simulation of the Transonic DFVLR-F5 Wing Experiment (W. Kordulla, Ed.)

Volume 23 Robust Multi-Grid Methods (W. Hackbusch, Ed.)

Volume 26 Numerical Solution of Compressible Euler Flows (A. Dervieux / B. van Leer / J. Periaux / A. Rizzi, Eds.)

Volume 27 Numerical Simulation of Oscillatory Convection in Low-Pr Fluids (B. Roux, Ed.)

Volume 28 Vortical Solution of the Conical Euler Equations (K. G. Powell)

Volume 29 Proceedings of the Eighth GAMM-Conference on Numerical Methods in Fluid Mechanics (P. Wesseling, Ed.)

Volume 30 Numerical Treatment of the Navier-Stokes Equations (W. Hackbusch / R. Rannacher, Eds.)

Volume 31 Parallel Algorithms for Partial Differential Equations (W. Hackbusch, Ed.)

Volume 32 Adaptive Finite Element Solution Algorithm for the Euler Equations (R. A. Shapiro)

Volume 33 Numerical Techniques for Boundary Element Methods (W. Hackbusch, Ed.)

Volume 34 Numerical Solutions of the Euler Equations for Steady Flow Problems (A. Eberle / A. Rizzi / E. H. Hirschel)

Volume 35 Proceedings of the Ninth GAMM-Conference on Numerical Methods in Fluid Mechanics (J. B. Bos / A. Rizzi / I. L. Ryhming, Eds.)

Volume 36 Numerical Simulation of 3-D Incompressible Unsteady Viscous Laminar Flows (M. Deville / T.-H. Lê / Y. Morchoisne, Eds.)

Volume 37 Supercomputers and Their Performance in Computational Fluid Mechanics (K. Fujii, Ed.)

Volume 38 Flow Simulation with High-Performance Computers I (E. H. Hirschel, Ed.)

Volumes 1 to 7, 9 to 11, 13, 15, 16, 19 and 21 are out of print.

Supercomputers and Their Performance in Computational Fluid Dynamics

Edited by
Kozo Fujii

vieweg

Die Deutsche Bibliothek – CIP-Einheitsaufnahme

Supercomputers and their performance in computational fluid dynamics / ed. by Kozo Fujii. – Braunschweig; Wiesbaden: Vieweg, 1993
(Notes on numerical fluid mechanics; Vol. 37)

NE: Fujii, Kozo [Hrsg.]; GT

Softcover reprint of the hardcover 1st edition 1993
Vieweg ist a subsidiary company of the Bertelsmann Publishing Group International.

Produced by W. Langelüddecke, Braunschweig
Printed on acid-free paper

ISSN 0179-9614
ISBN-13: 978-3-528-07637-5 e-ISBN-13: 978-3-322-87863-2
DOI: 10.1007/978-3-322-87863-2

PREFACE

Supercomputer technologies have evolved rapidly since the first commercial-based supercomputer, CRAY-1 was introduced in 1976. In early 1980's three Japanese supercomputers appeared, and Cray Research delivered the X-MP series. These machines including the later-announced CRAY-2 and NEC SX series created one generation of supercomputers, and the market was spread dramatically. The peak performance was higher than 1 GFLOPS and the compiler improvement was remarkable. There appeared many articles and books that described their architecture and their performance on several benchmark problems. The late 1980's saw a new generation of supercomputers. Following CRAY Y-MP and Hitachi S-820 delivered in 1988, NEC announced SX-3 and Fujitsu announced the VP2000 series in 1990. In addition, Cray Research announced the Y-MP C-90 late in 1991. The peak performance of these machines reached several to a few ten's GFLOPS. The hardware characteristics of these machines are known, but their practical performance has not been well documented so far.

Computational Fluid Dynamics (CFD) is one of the important research fields that have been progressing with the growth of supercomputers. Today's fluid dynamic research cannot be discussed without supercomputers and since CFD is one of the important users of supercomputers, future development of supercomputers has to take the requirements of CFD into account. There are many benchmark reports available today. However, they mostly use so called kernels. For fluid dynamics researchers, benchmark test on real fluid dynamic codes are necessary.

The scope of the present book is as follows. First, the features of new-generation supercomputers are reviewed. Their architectures and capabilities are described by people representing the supercomputer manufacturers. All the contributors are the key persons who have been engaged in developing leading-edge supercomputers. Dr. Kent Misegades, who is a specialist of application softwares in Cray Research was scheduled to write the article for Cray, but unfortunately left the company last year. Therefore, Mr. Toshihiro Hongo at Cray Japan wrote the article on the CRAY Y-MP C-90 instead. Second, a benchmark test result using a realistic CFD computer code is reported. A series of benchmark tests using a real CFD Navier-Stokes code has been carried out since 1989 as a collaborative work between the Office of Naval Research in the United States and the Institute of Space and Astronautical Science in Japan. The final result is included in this book, together with future requirements on supercomputer performance. Supercomputer development should be discussed from many point of views. Three prominent researchers discuss the feature of new-generation supercomputers from their own view points. Dr. Kenneth W. Neves discusses the hardware development in Chapter V. Dr. Wolfgang Gentzsch discusses the vectorization and parallelization techniques utilizing these new machines. Mr. Hajime Miyoshi et al. review the requirements of CFD for supercomputer performance and describe architecture of the machine that realizes these requirements. It should be noted that Mr. Miyoshi, although he is not well known outside of Japan, is one of the key persons in the development of all supercomputers in Japan. These chapters improve our understanding of new-generation supercomputers and balance the chapters on hardware architecture. I hope, the book gives readers a better understanding of new-generation supercomputers and becomes useful to the CFD researcher for their efficient use.

I wish to thank the chief editor of this series, Prof. E. H. Hirschel who suggested to make this book, and the other editors of this series for the valuable discussions and suggestions concerning this volume. I wish to thank all the contributors. The effort of all the people associated with our benchmark test are greatly appreciated. Without their effort, the book would not have been realized. Finally, I would like to thank my secretary, Chiho Saito as well as the members in my laboratory, especially one of the graduate students, Fumio Shimizu for setting up the tex format and putting all the manuscripts into this same format.

June 1992 K. Fujii

CONTENTS

I CRAY Y-MP C90 SUPERCOMPUTER

Toshihiro Hongo
Cray Research Japan, Co., Ltd.
Ichibancho Eight-One Bldg., 3rd Floor,
6-4, Ichibancho, Chiyoda-ku, Tokyo, 102 JAPAN

1. INTRODUCING THE CRAY Y-MP C90 SUPERCOMPUTER

In 1822, an Englishman named Charles Babbage invented the Difference Engine, the forerunner to modern computers. As the technical wonder of its age, it revolutionized the way people thought about problem solving. Now, Cray Research starts a new revolution in science and engineering with the CRAY Y-MP C90 supercomputer, a new Difference Engine for the Grand Challenges of today.

The CRAY Y-MP C90 supercomputer is a bold new engineering achievement that offers unmatched problem-solving capabilities. As the most powerful supercomputer ever offered by Cray Research, the CRAY Y-MP C90 uses many innovative technologies to deliver easy access to the highest levels of performance and reliability.

At Cray Research, we offer more than just the fastest hardware and most comprehensive software, we offer a Network Supercomputing solution that allows you to rise to the challenges of your work. As a complete system, Cray Research hardware, software, applications and networks provide the insight you need to approach today's problems with revolutionary solutions.

2. REDEFINING HIGH-PERFORMANCE COMPUTING

As the most powerful supercomputer available for production applications, the CRAY Y-MP C90 supercomputer is a major milestone of technology leadership. It sets new standards for high performance computing with six times the peak computing power of the original CRAY Y-MP 8 system, while maintaining upward binary compatibility.

The CRAY Y-MP C90 system features an all new central processor that provides a theoretical peak performance of 1.5 billion floating-point operations per second (1.5 GFLOPS), a substantial peak performance of 1.0 GFLOPS. To achieve this sustained performance, each CPU delivers four vector results in each clock period with a dual vector pipeline. Using 16 of these powerful processors and 256 million words (2 billion bytes) of central memory, the CRAY Y-MP C90 system enables a higher level of productivity for scientists and engineers with a theoretical peak performance of 24 GFLOPS.

Peak performance alone does not ensure computational efficiency; to sustain high levels of performance, a supercomputer requires a balanced system of hardware and software where all parts are matched to provide the most useful power to the user. The balanced architecture of the CRAY Y-MP C90 system includes the following elements to ensure maximum performance:

* More parallelism than any other vector supercomputer available today. The 16-processor CRAY Y-MP C90 features 64-way vector parallelism with two vector pipes and two functional units for each vector pipe per CPU. Combined with its mature, production-tested multitasking software, the CRAY Y-MP C90 system makes it easy to apply the power of 64-way vector parallelism to today's most widely used science and engineering codes.
* Unprecedented memory bandwidth. The CRAY Y-MP C90 delivers more sustained computing power to the user with four memory ports per CPU and 250 Gbytes/sec of memory bandwidth.
* Unmatched I/O bandwidth. To run efficiently, a high-speed supercomputer requires equally expansive input/output capabilities. With an aggregate I/O bandwidth of 13.6 Gbytes/sec, the CRAY Y-MP C90 system offers the most powerful and versatile I/O capabilities in the industry.
* Optional SSD Solid-state Storage Device for increased throughput. An optional SSD Solid-state Storage Device provides very-high-speed secondary memory with up to 2 Gwords (16 Gbytes) of storage capacity. With an aggregate bandwidth of 13.6 Gbytes/sec, the SSD allows you to solve large problems in less time.

3. BRIDGING THE GAP BETWEEN POTENTIAL AND PRODUCTIVITY

In addition to solving problems much larger than previously possible, the CRAY Y-MP C90 system can solve a greater number of problems as well. But increased productivity doesn't stop there. All Cray Research systems feature a powerful software application support environment that eliminates obstacles between you and the problems you need to solve. The application support environment offers the following advantages unique to Cray Research systems:

* Maximized throughput increases user productivity. The CRAY Y-MP C90 delivers the maximum amount of useful power to the user with UNICOS, the most robust and production-capable UNIX-based operating system available.
* Proven application development environment. UNICOS allows you to develop complex applications quickly and easily with a broad range of powerful compilers, debuggers, libraries, visualization capabilities, and distributed processing tools.
* High-speed I/O increases performance. The I/O capabilities of the CRAY Y-MP C90 system are enhanced further with powerful I/O software that provides exceptional throughput automatically. This software also provides simple-to-use options for reducing I/O transfers and attaining peak disk and tape transfer rates.
* Network supercomputing delivers real supercomputing power to your desktop. Because Cray Research supercomputers support industry standards as well as a variety of language extensions, network protocols, and utilities from other vendors, they can be integrated easily into heterogeneous computing environments.
* A proven set of powerful applications. Cray Research offers leading-edge, optimized applications for nearly every scientific and engineering discipline, including the most widely used third-party application programs.

4. PROTECTING YOUR HIGH-END SUPERCOMPUTING INVESTMENTS

To protect your high-end supercomputing investments, the CRAY Y-MP C90 offers full upward binary compatibility with the entire CRAY Y-MP family of supercomputers. Applications developed on any CRAY Y-MP system can be run on the CRAY

Y-MP C90 system.

The proven technologies of the CRAY Y-MP family of supercomputers offer the widest range of computational capabilities available. From air-cooled CRAY Y-MP EL systems to full-scale capability supercomputers like the CRAY Y-MP 8 and CRAY Y-MP C90 systems, the CRAY Y-MP family features binary-compatible performance spanning more than two orders of magnitude.

The CRAY Y-MP C90 provides an easy upgrade path from other CRAY Y-MP systems. Customers wishing to upgrade from their existing CRAY Y-MP 8E system can do so by simply swapping mainframe cabinets. The CRAY Y-MP C90 system can use existing IOS, SSD, and support equipment, easing system installation and providing significant cost savings.

5. THE BEST OVERALL SUPERCOMPUTING SOLUTIONS

To enhance productivity, all CRAY Y-MP systems can process a varied workload mix and still maintain optimal performance. While some architectures offer fast solutions for only certain types of problems, the balanced CRAY Y-MP architecture offers the highest possible performance on scalar, short vector, long vector, parallel and highly parallel problems. Because the real problems making up a production workload consist of a varied mix of codes, the CRAY Y-MP C90 system offers the best overall performance solution.

6. NEW TECHNOLOGIES MAXIMIZE SYSTEM AVAILABILITY

To deliver unmatched performance and reliability, the CRAY Y-MP C90 system uses the following new technologies:

* Custom high-speed 10000 gate array circuits increase reliability and reduce manufacturing costs with four times the level of integration of previous devices.
* Surface-mount component assembly reduces manufacturing and reliability problems associated with chip leads.
* Multi-layer circuit boards with internal pathways prevent contaminants from corrupting signal integrity.

7. PHYSICAL DESCRIPTION

The CRAY Y-MP C90 system is comprised of a mainframe cabinet and IOS cabinet. The mainframe cabinet contains the CPUs, central memory and power supplies. The IOS cabinet contains the I/O subsystem and optional SSD. The mainframe cabinet measures 116.25 inches wide by 101 inches deep by 85.5 inches high (2.95 m × 2.57 m × 2.17 m). The IOS cabinet measures 46 inches wide by 75.5 inches deep by 76.25 inches high (1.17 m × 1.92 m × 1.94 m).

8. CRAY Y-MP C90 HIGHLIGHTS

* Fully upward compatibility with entire CRAY Y-MP line
* Sixteen processors
* 24 GFLOPS theoretical peak performance
* 16 GFLOPS sustained peak performance
* 10000-gate Very Large Scale Integration (VLSI) gate-array circuits
* Dual vector pipelines in each CPU
* Flexible hardware chaining for vector operations
* Gather/scatter and compressed index vector support
* Flexible processor clustering for multitasking applications
* Four parallel memory ports per processor
* Liquid cooling system
* Dedicated registers for efficient interprocessor communication and control
* 256 Mwords (2 Gbytes) central memory
* Single-byte correction, double-byte detection memory protection
* Memory bandwidth of 250 Gbytes/sec
* Optional SSD with 512, 1024 or 2048 Mwords (4, 8 or 16 Gbytes)

9. THE MOST POWERFUL I/O TECHNOLOGY AVAILABLE

The IOS allows the central memory of the CRAY Y-MP C90 system to communicate at high speeds with networks and peripherals such as disk and tape drives. To provide industry-leading performance, versatility, and reliability, the CRAY Y-MP C90 system uses the Model E I/O Subsystem (IOS) recently introduced with the CRAY Y-MP systems.

To provide the CRAY Y-MP C90 system with massive production workload capacity, the I/O architecture delivers up to 13.6 Gbytes/sec of I/O bandwidth. This very large bandwidth allows you to connect to more peripheral devices and perform more simultaneous activities.

The IOS provides a flexible framework that can grow with your I/O and peripheral needs. The IOS is comprised of two to 16 I/O Clusters (IOCs), which allow large amounts of data to be accessed with support for high performance data transmission. Each IOC supports up to 16 channel adapters for a total of 256 channels for connection to disk storage units, tape units, and communications connections.

The standard configuration of the CRAY Y-MP C90 system includes two I/O clusters and 16 channel adapters. Additional I/O clusters and channel adapters can be configured easily at your site.

10. INPUT/OUTPUT HIGHLIGHTS

* High performance I/O software
* 12800 MIPS I/O subsystem
* Up to 16 I/O clusters
* Up to 16 channel adapters per cluster for a total of 256 channels
* Aggregate I/O bandwidth of 13.6 Gbytes/sec
* Support for high-performance disk drives
* Support for high-performance on-line tapes
* Support for industry-standard and proprietary networks
* Support for the ANSI standard HIPPI channel

11. ADVANCED SSD TECHNOLOGY

The optional SSD Solid-state Storage Device is a very fast random-access device that increases system performance when used in the following ways:

* As high-speed I/O cache (ldcache).
* As secondary program memory (sds) supported by libraries and the CF77 Fortran compiling system.
* As a system swap space.

The CRAY Y-MP C90 system uses the proven Model E SSD technology that allows you to improve system throughput and increase individual job performance with an aggregate bandwidth of up to 13.6 Gbytes/sec. Using VLSI chips and increased system integration, the SSD is available with up to 2 Gwords (16 Gbytes) of storage capacity.

The CRAY Y-MP C90 system communicates with the SSD through up to four 1800-Mbytes/sec channels. The SSD is connected to the IOS through 200-Mbytes/sec channels. These connections enable data to be transferred directly between an IOS and the SSD without passing through central memory, thereby increasing overall performance.

12. SSD HIGHLIGHTS

* Improves system throughput and individual program performance
* Up to 2 Gwords (16 Gbytes) capacity
* Reliable VLSI technology
* Up to four 1800-Mbytes/sec channels to mainframe
* Aggregate SSD bandwidth of up to 13.6 Gbytes/sec
* SECDED memory protection
* 200-Mbytes/sec channels to the IOS

13. DISK DRIVES

Cray Research offers fast, reliable mass storage devices that provide large storage capacities in a small physical space. The CRAY Y-MP C90 system supports all current Cray Research disk storage devices including the DD-60 and DD-61 disk storage units. With the capability to support over 4 terabytes of disk storage, the CRAY Y-MP C90 gives you high-speed access to 18 times more on-line data than was possible with the original CRAY Y-MP Model D system.

The DD-60 disk drive offers outstanding performance and large storage capacities when matched with the I/O capability of the CRAY Y-MP C90 system and the UNICOS operating system. Each I/O cluster configured with 16 DD-60 disk drives can deliver up to 320-Mbytes/sec performance (up to 200 Mbytes/sec in each direction).

The DD-60 is a 24-Mbytes/sec disk drive with a sustained transfer rate of 20 Mbytes/sec and a capacity of 1.96 Gbytes. Disk striping provides single disk transfer rates of several times the rate of individual disk drives. Up to eight DD-60 disk drives can be connected to each disk channel adapter in the IOS.

The DD-61 disk drive delivers large storage capacities at a low cost. The DD-61 provides you with access to large amounts of data using highly reliable, 8-inch disk technology that gives the DD-61 a lower cost per Mbyte, a small footprint, and low power consumption. The DD-61 is a 3-Mbytes/sec disk drive with a sustained transfer rate of 2.6 Mbytes/sec and a capacity of 2.23 Gbytes. Up to eight DD-61 disk drives can be connected to each disk channel adapter in the integrated I/O subsystem.

14. SOFTWARE

14.1 Performance Oriented, Feature-Rich Software

The Cray Research application support environment is a complete body of performance-oriented, standards-based system software that enables users to focus on their work, not the system's requirements. As part of a total system solution, the application support environment includes UNICOS, the world's first and highest-performance UNIX-based supercomputer operating system, as well as a set of powerful compilers, development tools, high-performance libraries and data storage systems.

14.2 UNICOS Operating System

The UNICOS operating system is the most powerful and feature-rich UNIX-based operating system available to supercomputer users. Based on the UNIX System V operating system with Berkeley extensions and numerous performance enhancements, UNICOS is an interactive and batch operating system that offers a number of advantages including high performance, full functionality, standards and connectivity.

UNICOS is the result of hundreds of programmer years of optimizations that deliver very high performance on production workloads. Together with the powerful CRAY Y-MP C90 computer hardware and autotasking capabilities, this performance not only provides fast turn-around on individual jobs, but also high throughput for a varied workload through sophisticated job scheduling capabilities.

UNICOS combines all the inherent strengths of UNIX, such as a familiar user interface, with production-oriented features including high-performance I/O, optimal memory bandwidth utilization, multiprocessing support, ANSI/IBM tape support, resource control, sophisticated job scheduling, tunable accounting, multi-level security and batch processing.

To help programmers build powerful applications in less time, UNICOS includes a powerful application development toolset. Visual and line-oriented tools automatically analyze codes for programming and performance problems, offer expert advice for making improvements, and offer graphical interfaces to speed assimilation and manipulation of this information.

14.3 UNICOS Highlights

Full production functionality

- Batch and interactive processing
- High-speed tape support
- Resource management
- Extensive accounting features
- Checkpoint/restart
- Networking
- Data Migration Facility (DMF)
- On-line system diagnostics
- IEEE data format support
- Multi-level security

High performance

- Autotasking and autovectorizing features
- Efficient, asynchronous I/O
- Striped disks
- File system extensions

Ease of use

- Industry standard interfaces
- Compatibility across entire CRAY Y-MP product line
- Advanced program development tools
- X window system support

- Performance analysis tools

14.4 Compilers

Cray Research offers the most powerful compilers in the industry, including the CF77 Fortran compiling system, the Cray standard C compiler, Cray Ada and Pascal. The CF77 compiling system was the first Fortran compiler in the industry with support for automatic parallel processing, automatic vectorization and scalar optimization. These compiling features typically require little or no code modification by the user, and full optimization is on by default.

The CF77 5.0 compiling system ensures portability with full validated compliance to ANSI standard 3.9-1978. The flexibility of CF77 allows it to accept many nonstandard constructs written for compilers from other vendors.

The CF77 compiling system compiles Fortran77 programs into executable code modules that take full advantage of the CRAY Y-MP C90 vector capabilities, while its autotasking feature further enhances performance. For those codes that are not highly vectorizable, CF77 ensures the best possible execution time by providing maximized scalar optimization for the CRAY Y-MP C90 system.

Because supercomputing applications written in C language are becoming increasingly popular, Cray Research offers the highest-performance validated ANSI standard C compiler in the industry. The Cray standard C compiler can be used to create portable, highly optimized code with performance comparable to Fortran programs. Like CF77, the Cray Standard C release 3.0 compiler takes full advantage of the CRAY Y-MP C90 performance capabilities with automatic vectorization, scalar optimization and autotasking.

14.5 Autotasking

The CF77 compiling system and the Cray standard C compiler include autotasking features that can dramatically improve performance on all CRAY Y-MP systems. The autotasking feature divides a program into discrete tasks that can be performed concurrently on multiple processors in the CRAY Y-MP C90 system. The autotasking features also include a convenient, powerful set of directives that allow programmers to fine-tune their code for even better performance. In production environments, this feature can be used to improve both individual job performance and overall system throughput.

14.6 UNICOS Storage System

The UNICOS storage system is the world's first high-performance UNIX-based file server. With the UNICOS storage system, Cray Research systems enable users to meet their computing needs while also addressing the file storage needs of their network. The UNICOS storage system provides transparent data access, file access capabilities, system administration, and automated storage management capabilities.

14.7 Applications

Cray Research offers leading-edge applications for nearly every scientific and engineering discipline, including the most widely used third-party application programs. These applications are used by diverse industries to accelerate product development, increase productivity, and solve basic research problems. Applications are available for industries such as aerospace, automotive, electronics, chemistry, energy, petroleum and defense.

To augment the utility of its supercomputers, Cray Research has developed specialized, key application packages such as:

* UNICHEM, Cray Research's easy-to-use supercomputing environment for computational chemistry simulation that enables researchers to explore complex chemical systems at a new level of detail from their desktops.
* CRI/TurboKiva, a powerful combustion simulation environment that provides a reliable and effective analysis tool for designing cleaner-burning internal combustion engines. CRI/TurboKiva is an enhanced engineering version of Kiva II, Los Alamos National Laboratory's public domain engine flow, spray, and chemical reaction simulation code.

14.8 The Power of Visualization

Cray Research supports the following software packages to couple the power of visualization with its supercomputers:

* Multipurpose Graphic System (MPGS), an interactive menu-driven engineering visualization package for use on Cray Research computer systems. MPGS works with a wide variety of engineering applications.
* AVS (Application Visualization System), a high-level, object-oriented environment for constructing visualization programs.
* IRIS Explorer, a visually based, object-oriented application building environment.
* The Cray Visualization Toolkit (CVT), which enables users to run applications on Cray Research systems through their workstations. CVT allows users to generate graphics and graphical user interfaces easily with the following tools:
  - Release X11R4 of the X window system
  - Sun Microsystems' XView toolkit (OPEN LOOK)
  - Open Software Foundation's (OSF) motif 1.1 toolkit
  - Silicon Graphics, Inc. Distributed Graphics Library (DGL)

These tools allow most applications that run on Cray Research systems to have the same "look and feel" as the most common workstation environments, making Cray Research systems even easier to use and making users more productive.

15. NETWORK SUPERCOMPUTING

15.1 Delivering Supercomputing Power to Your Desktop

To bring the benefits of supercomputing to more users than ever before, Cray Research is dedicated to making its systems accessible through network supercomputing. Because Cray Research supercomputers support industry standards as well as a variety of language extensions, network protocols and utilities from other vendors, they can be integrated easily into heterogeneous computing environments.

An array of communication products and protocols supported by Cray Research allows applications to be distributed within your network. Through the implementation of official, emerging and de facto networking standards, Cray Research provides connectivity to most UNIX-based mainframes, minicomputers and workstations. These standards include the TCP/IP networking protocol and applications, the X window system, the network file system, the Open Systems Interconnection (OSI) of the International Standards Organization (ISO), the High Performance Parallel Interface (HIPPI), the

Fiber Distributed Data Interface (FDDI), as well as other networking standards.

Cray Research also provides network batch software products that offer access to proprietary protocol implementations (such as SNA, DECnet, and CDCNET) through network gateways. Network batch software runs on a variety of systems and workstations to provide the logical connection to a CRAY Y-MP computer system. Standard network batch software is available for the following systems: IBM MVS and VM, CDC NOS, NOS/VE, DEC VAX/VMS and a variety of computers and workstations running the UNIX operating system. Network batch software for Unisys and Honeywell Bull systems is available from third-party vendors.

Network supercomputing increases user productivity by allowing access to a wide range of computing platforms for optimal workload distribution. The result is a combination of flexibility and computing power unparalleled in the computer industry.

16. SUPPORTABILITY

16.1 Maximized System Availability

The CRAY Y-MP C90 supercomputer uses new technologies to provide high system reliability while maintaining high performance. It includes the following design enhancements that increase its serviceability:

* Graceful degradation of CPUs allows CPUs to be configured in and out of the UNICOS operating system without interrupting system operation.
* CPU isolation enables diagnosis of isolated CPUs while other CPUs run the operating system and applications.
* Connectorized signal, power and cooling connections between modules and the mainframe chassis allow for efficient removal and replacement.

Cray Research offers a wide range of maintenance options for the CRAY Y-MP C90 system to meet your needs. To assure high system availability, Cray Research has developed advanced system support tools including the new System Maintenance and Remote Test Environment (SMARTE), which provides continuous error detection and isolation. SMARTE is a UNIX-based system that schedules diagnostic activity, automatically reports errors, and provides a common X window system interface to all on-line diagnostic and concurrent maintenance tools.

17. THE CRAY Y-MP C90 SUPERCOMPUTER, NOTHING ELSE COMES CLOSE

The CRAY Y-MP C90 system is the most powerful supercomputer available for technical computing. As part of a complete supercomputing solution, it provides the insight you need to approach today's production problems and Grand Challenges with revolutionary solutions. CRAY, CRAY Y-MP, SSD and UNICOS are registered trademarks, and autotasking, CF77, CRAY Y-MP EL, CRI/TurboKiva, DD-60, DD-61, IOS, MPGS, SMARTE and UNICHEM are trademarks of Cray Research, Inc.

AVS is a trademark of Stardent Computer, Inc. CDC, CDCnet, NOS/BE and NOS/VE are trademarks of Control Data Corporation. Data General is a trademark of Data General Corporation. DEC, DECnet, VAX, VAXcluster and VMS are trademarks of Digital Equipment Corporation. Ethernet is a trademark of the Xerox Corporation. IBM, MVS and VM are trademarks of International Business Machines Corporation. IRIS Explorer is a trademark of Silicon Graphics, Inc. LANlord is a trademark of Computer Network Technology Corporation. Motorola is a trademark of Motorola, Inc. Sun Workstation is a trademark of Sun Microsystems, Inc. The Cray Research implementation of TCP/IP is based on a product from the Wollongong Group, Inc. UNIX is a trademark of UNIX System Laboratories, Inc. The X window system is a trademark of the Massachusetts Institute of Technology.

Table 1 CRAY Y-MP C90/16256 Product Specifications

CPU	
Technology	Custom silicon 10,000 gate array circuits
Number of CPUs	16
Vector pipes	2 per CPU
Theoretical Peak performance	24 GFLOPS
Sustained Peak Performance	16 GFLOPS
Memory	
Memory ports	4 double-word ports per CPU
Technology	15 ns BiCMOS
Memory size	256 Mwords (2 Gbytes)
Total memory bandwidth	250 Gbytes/sec
I/O	
Number of I/O clusters	2 - 16
Total system I/O bandwidth	13.6 Gbytes/sec
HIPPI channels	100 Mbytes/sec
LOSP channels	6 or 20 Mbytes/sec
HISP channels	200 Mbytes/sec
VHISP channels	1800 Mbytes/sec
Optional SSD	
Capacity	512, 1024, 2048 Mwords (4, 8, or 16 Gbytes)
Bandwidth	13.6 Gbytes/sec
Physical characteristics	
Mainframe cabinet footprint area	81 ft^2 (7.50 m^2)
IOS cabinet footprint area	23.9 ft^2 (2.20 m^2)

Table 2 CRAY Y-MP C90 System Configuration

Model	CPUs	Central Memory	I/O clusters	Optional SSD
CRAY Y-MP C90/16256	16	256 Mwords (2 Gbytes)	2 - 16	512, 1024, or 2048 Mwords (4, 8, or 16 Gbytes)

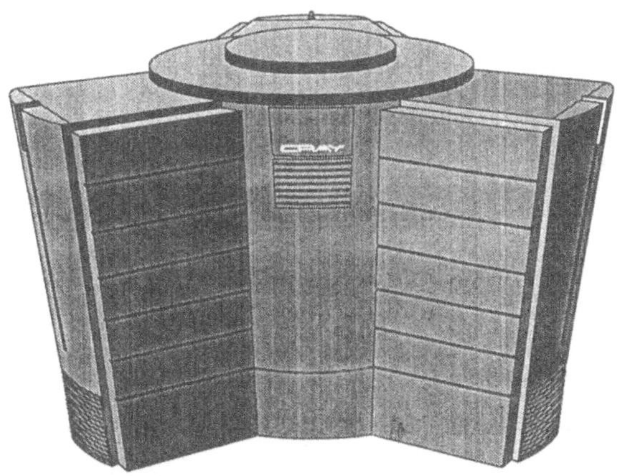

Fig. 1 View of Computer

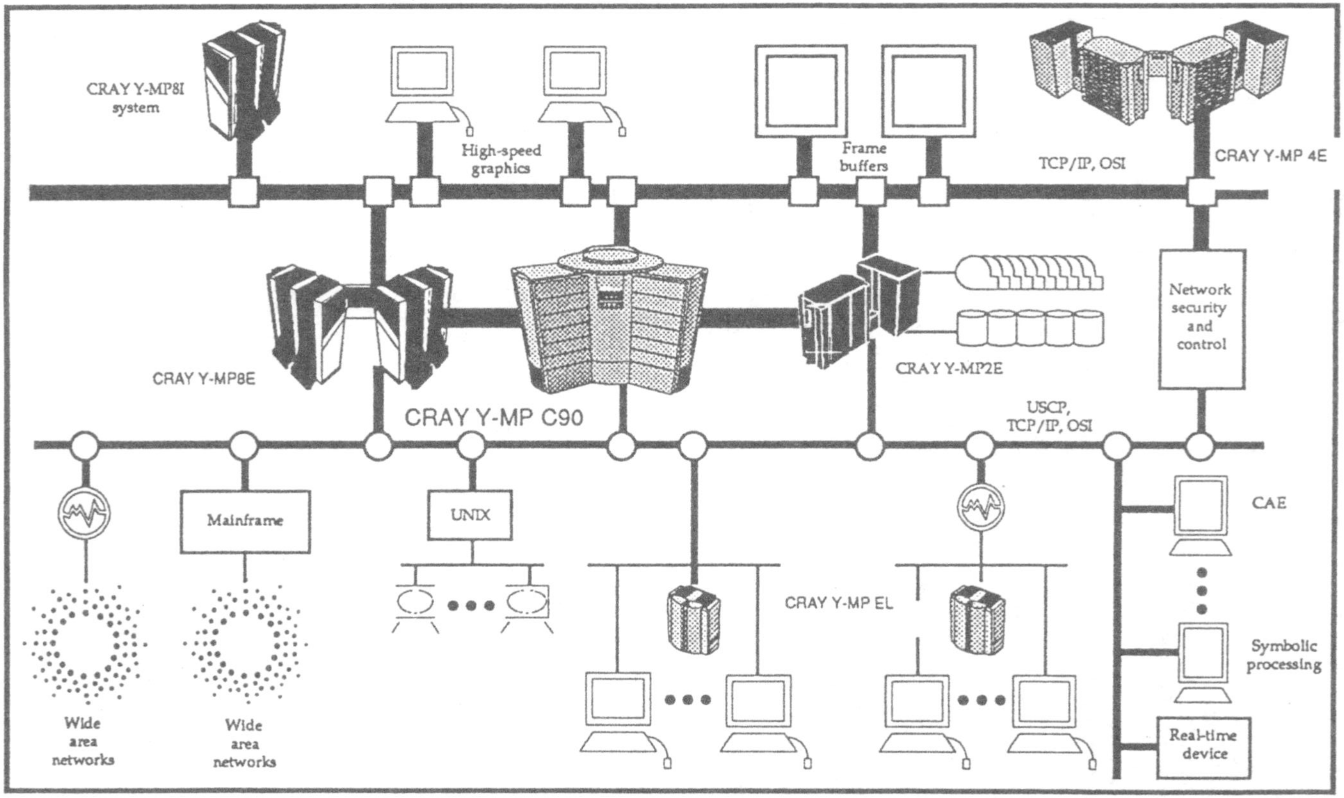

Fig. 2 Network Capabilities

II FUJITSU VP2000 SERIES SUPERCOMPUTER

Keiichiro Uchida
Supercomputer Development Department, Main Frame Division,
Computer Systems Group, Fujitsu Limited,
1015, Kamikodanaka, Nakahara-ku, Kawasaki, 211 JAPAN

SUMMARY

Fujitsu VP2000 series Supercomputer is a pipeline supercomputer series developed by Fujitsu. The peak performance of this series ranges from 375 MFLOPS of the VP2100/10S to 5 GFLOPS of the VP2600/10. Operations are mainly performed by using a pipeline which is called "Universal Pipeline". This pipeline provides a capability of one of the following operations such as a multiplication, an addition, a multiplication-addition or a logic operation at one time. In addition, in case that two pipelines are used, they can perform such operations concurrently. The VP2200, the VP2400 and the VP2600 have two pipelines and both pipelines are used for multiplication-addition concurrently to attain the peak performance. FORTRAN77 EX/VP, which is developed to utilize this hardware, makes it possible to schedule these two pipelines effectively. Using the VP2600/10, the performances of LINPACK problem for a matrix of order 100 and for a matrix of order 1000 are 249 MFLOPS and 4009 MFLOPS respectively.

1. INTRODUCTION

In 1982, Fujitsu announced its first supercomputer the VP-100/200[1]. The machine was based on Fujitsu's previous experience with the FACOM 230-75 Array Processor (APU)[2] and the technology of the M380 which was one of the largest scalar computer at that time. Each of its theoretical peak performance is 267 and 533 MFLOPS respectively. In 1985, the VP-400 and the VP-50 were announced for the demand of high performance and low price machine. In 1986, the VP-30 was announced as the lowest price machine. The VP-400, the VP-50, and the VP-30 were with peak performance of 1.14 GFLOPS, 140 MFLOPS and 110 MFLOPS respectively.

The VP series E model[3] which employed a multiply & add pipeline was announced in 1987. Each of the peak performance of the VP-30E and the VP-50E was improved by 2 times to that of the VP-30 and the VP-50. Each of the peak performance of the VP-100E, the VP-200E, and the VP-400E was improved by 1.5 times to that of the VP-100, the VP-200 and the VP-400.

The VP2000 series[4-6] was announced in 1988 is the succession to the VP series E model.

The series consists of the VP2600 (high-end model), the VP2400, the VP2200, the VP2100 and the VP2100S (low-end model). In addition, the series can be classified into three types according to the number of scalar processors, such as the Uni-processor (model 10), Dual-scalar processor (model 20), and Quadruple-scalar processor (model 40). Figure 1 shows these models. In total, there are 11 models which can all be upgraded in the fields (Fig. 2).

Table 1 summarizes the specifications of the VP2000 series. The high-end model the VP2600 (a picture is shown as Fig. 3) can offer 5 GFLOPS (5000 million of 64 bit-FLoating-point Operations Per Second) which is 13 times faster than the entry-level model (VP2100S) in maximum performance. Figure 4 shows a block diagram of the Fujitsu VP2000 series. The system consists of vector processing unit (VPU), main storage unit (MSU), and channel processor (CHP). The VPU corresponds to the central processing unit (CPU) of general-purpose computer, and consists of scalar unit (SU) and vector unit (VU).

The new features supported in the VP2000 series are the adoption of a multiprocessing architecture, the expansion of main storage, the introduction of system storage, and connection to UNIX* environments. Employing the architecture and the technology, Fujitsu VP2000 series seeks high performance, expandability, flexibility, and an open system.

* UNIX is a registered trademark of UNIX System Laboratories, Inc.

2. ARCHITECTURE

The base architecture of the Fujitsu VP2000 series is the same with the general purpose computer, M series. In addition, the architecture of VP2000 series includes vector processing feature. The vector processing feature includes vector registers, mask registers, vector control registers and an address translation register.

2.1 Scalar Unit (SU)

The Fujitsu VP2000 series has 231 scalar instructions which are the same as Fujitsu's general-purpose computers, and includes the IBM 370 architecture.

The SU controls instruction execution by fetching and decoding instructions. Scalar instructions are executed in the SU, and vector instructions are issued to the VU. The SU contains 128 Kbytes of buffer storage, 16 general-purpose registers, 8 floating-point registers, and a scalar execution unit. Actual scientific and engineering calculations usually include many scalar operations in addition to vector operations. The VP2000 series realizes high scalar performance by providing powerful scalar processing capability.

2.2 Vector Unit (VU)

Vector instructions issued from the SU are executed in VU which consists of a large-capacity vector register and vector pipelines.

The VU uses 132 vector instructions (Table 2). The formats of these instructions are shown in Fig. 5. They are classified into two formats, VR type and VX type, each of which is six bytes long. Operations are executed in series for the amount of elements specified by the vector length (VL) in the vector control register.

The VR type instructions specify mainly the register-to-register operations such as operations between vector registers. The VR instructions have three operand fields. In general, operations are done between the two operands specified by R2 and R3, and the results are stored in a register specified by R1. Conditional vector operations (masked operation) are available in the VR instructions specifying mask register by R4 field.

The VX type instructions specify load/store operations. They are used to load vector or mask data from the MSU to R1 and to store vector or mask data from R1 to the MSU. The starting address of the location in the MSU where the elements are read or to be stored is specified by X2, B2 and D2. The distances between the addresses of consecutive elements are indicated in a general-purpose register specified by R2. R2 can be used for indirect addressing when it specifies a vector register. A masked operation is also available in the VX instructions using specific load/store instructions which are executed under a specific mask register.

The total of vector register capacity of the high-end model VP2600 is 128 Kbytes in which can store 16384 elements of eight-byte data. The configurations of the concatenated vector registers (number of elements by number of registers) are variable as follows: 64 × 256, 128 × 128, 256 × 64, ······, 2048 × 8. Mask registers have the total capacity of 2 Kbytes (element size is 1bit), and have the same configurations as the vector registers. (The configurations of registers are automatically defined by compiler to derive maximum performance from a source program.)

2.3 Main Storage Unit (MSU)

The VP2000 series has the MSU of maximum capacity 2 Gbytes which is accessed by the SU, the VU, and the CHP. The VU frequently accesses to the MSU and requires extremely large throughput to balance the processing speed of arithmetic execution pipeline. Therefore, a large number of data buses and bank interleaves (up to 512 ways) are provided. The VU can access the data stored in the MSU in contiguous, constant stride, and indirect addressing.

2.4 System Storage Unit (SSU)

The SSU is newly introduced to the VP2000 series which has the maximum capacity of up to 32 Gbytes, and can be used as the swapping area for jobs and virtual filing area for I/Os. Data can be transferred between the SSU and the MSU with the total maximum throughput of 2 Gbytes/sec. There are two types of transfer instructions: synchronous and asynchronous.

In synchronous type instruction, the data transfer is synchronized with the execution of the instruction. This means the transfer begins with the start of the instruction and completes with termination of the instruction.

In asynchronous type instruction, the instructions are terminated without waiting the completion of the transfer. The completion of instruction is notified by an interruption. This feature allows the VPU to execute another operation in parallel with the data transfer.

2.5 Channel Processor (CHP)

There are four types of channels which are the 4.5 Mbytes/sec electrical, 9 Mbytes/sec optical, 36 Mbytes/sec High-speed-optical, and 100 Mbytes/sec HIgh Performance Parallel Interface (HIPPI*) channels. Total throughput of 2 Gbytes/sec is attained with the maximum configuration of 256 channels.

* HIPPI is an interface specification of ANSI standard.

2.5.1 High-speed optical channel

The 36 Mbytes/sec optical channel can be connected with a large capacity high speed disk device named F6490. F6490 uses a disk array method which is composed of ten disks (Fig. 6). Every byte of eight-byte data is distributed to eight of these 10 disks. The other two disks are used as a parity and a backup disk. If a fixed error occurs in one of the eight disks, the disk will be replaced with the backup one. This disk array method allows higher throughput and higher reliability compared with conventional type disks.

2.5.2 HIPPI channel

HIPPI channel with throughput of 100 Mbytes/sec can be connected to the UltraNet* which is a high speed multi-vender LAN. This allows high speed data transfer between the VP2000 series supercomputer and workstations under multi-vender environment.

* UltraNet is a registered trademark of Ultra Network Technologies, Inc.

3. HARDWARE IMPLEMENTATION

This section describes the hardware techniques utilized in the VP2000 series.

3.1 Vector Pipelines

There are four types of vector pipelines implemented in the VP2000 series which are the load/store, mask, multiply & add/logical execution, and divide pipelines (Fig. 4). The operation rate of the vector pipeline depends on the model which is controlled by hardware.

Two identical load/store pipelines are provided (only one pipeline for model VP2100) to transfer data between a vector register (a mask register) and the MSU, which can be operated concurrently for continuously addressed data. The mask pipelines are for executing arithmetic operations with mask data. Two mask pipelines can be operated in parallel.

Two identical multiply & add/logical pipelines which are called universal pipelines are adopted in the VP2000 series. (Each of the VP2100 and the VP2100S have one pipeline.) Each pipeline can execute either of the multiply, add, multiply & add, first order recurrence or logical operations. These two flexible pipelines can be usefully operated by executing operations concurrently under various combinations. The divide pipeline is provided for executing especially the divide operation. The divide operation is realized with its throughput being 1/3 of the add or multiply operation.

3.2 Parallel Processing

Figure 7 shows instructions of control pipelines in the VPU. A vector instruction is issued from the SU to the VU. The VU controls instruction execution by transmitting the instruction to the appropriate vector control pipeline corresponding to the execution type. In the VU, two load/store pipelines (one-pipeline is used for the VP2100 and VP2100S), two mask pipelines, and two out of three arithmetic operation pipelines (one out of two for the VP2100 and VP2100S) can operate concurrently. In this way, the SU and six out of the seven vector pipelines can operate concurrently.

Each instruction control pipeline in the VU consists of R, S and T stages. The R stage controls fetching of operand data and start of instruction execution, meanwhile the T stage controls storing of operand data and termination of instruction processing. The S stage is an intermediate stage between R and T. When a vector instruction completes the R stage, information required for instruction control is retained in the S stage until the instruction moves into the T stage. This allows the next instruction to be executed immediately in the R stage after the preceding instruction is issued (Fig. 8).

LINKAGE is a facility that allows two pipelines to be connected logically for continuous operation as if they were a single pipeline. By using this facility, a vector instruction can read the data of the vector register or the mask register stored by the preceding instruction without waiting for completion of the instruction. Figure 9 shows an example which the vector register stored by the VLD (Vector Load Long) instruction. The data in the store pipeline is read by the succeeding VMAD (Vector Multiply & Add Long) instruction in the universal pipeline. The VMAD instruction can be issued when the universal pipeline becomes available after the instruction is in the T stage. If no LINKAGE facility is provided, the VMAD instruction cannot be started until completion of the VLD.

3.3 Advanced Scalar Operation

The VP2000 series has a large capacity of buffer storage and some new features for high-speed scalar operation. One-cycle instruction pipeline is implemented in the SU, so that scalar instruction can be issued to scalar execution unit in every cycle. As for the continuous floating-point operations, its execution cycles are reduced by the overlapping of execution sequence.

3.4 Other Features for High Speed Processing

Some vector instructions generate the result into a scalar register. In this case, the vector instruction may refer to the scalar register in which preceding vector instruction stored. The data to be sent to the scalar register is bypassed to the next vector instruction without waiting for the completion of the preceding vector instruction. Vector instructions, which access to the MSU in a random order, sometimes access the same address successively. In this case, the two accesses to the same address are merged to one in order to reduce wasteful accesses.

The VP2000 series achieve high effective performance with these advanced features of hardware.

4. MULTIPROCESSOR SYSTEM[7]

There are two types of multiprocessor systems (DSP and QSP) which improve effective performance of the entire system.

4.1 Dual Scalar Processor (DSP)

DSP is a new architecture in a multiprocessor system to increase system performance by attaching another SU.

The utilization of a VU is an important problem for the ordinarily vectorized application programs. For example, a program with vectorization ratio of 90% makes a VU busy less than 50% of the total CPU time. Taking advantage of this character, an extra SU may be attached which shares the VU in order to increase system performance.

The programmable registers in VU are provided for each SU (Table 1). The software, which is executed on an ordinary multiprocessor system, can be executed on this hardware. Vector instructions issued by two SUs are selected alternatively by hardware (the SU0 and the SU1 each corresponds to the SU in DSP system).

It should be noted that the improvement factor of the system throughput of DSP over UP is 2 when all the jobs are scalar, and that this factor is a monotonously decreasing function of the vectorization ratio of the programs, ending up with 1 when all the jobs are 100% vectorized. As an example, in the case where the vector vs scalar speedup factor is 10, assuming no contention of VU usage,the improvement factor of the system throughput will be expected up to 2 even when two programs running on DSP are both having 91% vectorization ratio.

4.2 Quadruple Scalar Processor (QSP)

In the VP2200 and the VP2400 model, there is another type of multiprocessor system called QSP. The QSP system is configured as a multiprocessor with two DSP systems sharing the MSU by using four SUs and two VUs. In the QSP system, multitasking can be performed to be distributed to each task of a single job to the processors.

5. HARDWARE TECHNOLOGY

5.1 Advanced LSIs

The logic circuits are designed using high-speed, high-density emitter-coupled logic (ECL). Each ECL LSI contains 15000 logic gates with 70 psec basic delay time (Fig. 10). In another type of LSI, 64 Kbytes of memory cells with 1.6 nano-seconds access time are integrated on a chip together with 3500 logic gates. As for the MSU, 1 Mbytes static RAM chips with 35 nano-seconds access time are adopted. GaAs LSI with 1200 gates is also used for data bus driver in the SSU to attain a high transfer throughput between the SSU and the MSU.

5.2 High Density Packaging

Multi-layer glass ceramics (MLG) is introduced in the VP2000 series supercomputer to attain a high transmission speed (Fig. 11). The SU is packaged on a 24.5 cm × 24.5 cm glass ceramic board with 61 layers.

5.3 Cooling Technology

A high performance conductive cooling module (CCM) is introduced to remove the large amount of heat generated by LSIs. The CCM has a superior cooling capability and permits heat dissipation up to 30 W per LSI, and 4.6 kW per board. The interface is protected from humidity to increase the reliability of the module by hermetically sealing the contacts between the CCMs and the MLG.

6. MSP SYSTEM

MSP (Multi-dimension Support Product) system which is a Fujitsu's proprietary operating system offers high processing capability, expandability, and flexibility. Some highlights regarding the utilization of a large capacity and high speed System Storage in the MSP for VIO/F (Virtual Input Output/Facility) file function and a swap function are described in the followings.

6.1 System Storage Usage

6.1.1 High speed large scale virtual I/O

In MSP, a high-speed I/O can be attained by using the VIO/F files. These files can be allocated on the MSU or the SSU. Since the SSU can have up to 32 Gbytes capacity, large I/O data can be accessed with high speed. The VIO/F files on the SSU can be passed to the next job step. This function is the improvement of the VIO/F function provided in the previous VP series E model systems.

6.1.2 High speed swapping

Swap feature is realized by utilizing the high-speed access time to the SSU. Higher performance can be attained by using the SSU than the DASD. This allows VPTSS (utilization of the VP feature from TSS terminals) sessions to use memory space beyond the MSU capacity. An asynchronous transfer feature of the SSU is also fully utilized in implementing the swap function and the data transfer between the MSU and the SSU. This feature enables the SU to dedicate to its computations.

6.2 Support of DSP/QSP

As shown in Fig. 12, a DSP/QSP can be also configured to be an asymmetrical multiprocessor system. When the vectorization ratio is high, a VU should be dedicated to execute vector jobs. In this case, one SU dedicating the VU is dispatched the vector jobs and another SU is dispatched the scalar jobs. The asymmetrical configuration of DSP/QSP model is a way to utilize the hardware capability to its maximum, and is suitable for the computing center environment where many jobs with high vectorization ratio and many scalar jobs exist.

6.3 Virtual Machine

A new virtual machine function, AVM/EX (Advanced Virtual Machine/Extended) has been realized with small overhead and high reliability and efficiency, by enhancing the functions of a virtual machine. A large advantage which the AVM/EX offers is that the VP feature can be utilized simultaneously by multiple guest operating systems (four at maximum). Users at the workstation can select a suitable OS for their objectives.

6.4 TCP/IP Support

The MSP system supports the TCP/IP protocol, which is a standard for communications in the UNIX environment. MSP system enhances the TCP/IP protocol to connect the VP2000 series with UNIX work stations. This enables this series to utilize UNIX workstations and any other supercomputer without providing a gateway.

7. UNIX®* SYSTEM

A Fujitsu's UNIX system for mainframe computers, the UXP/M and VPO (Vector Processor support Option) is available on the VP2000 series and provides various functions in UNIX environment. The main functions of the VPO are described in the following.

7.1 Optimization of Vector Processes

The processes using the VU, called VP processes, are managed separately from normal processes. The VP processes usually require a large memory area and long CPU time. Therefore, the process scheduler controls a time slice value for the VP processes. This value can be changed by the system administrator while the system is running. A memory manager of the VPO permits processes using up to 2 Gbytes memory with 1 Mbytes page size.

7.2 High-Speed I/O Access

I/O access time must balance with operation speed of the vector processor. Under UXP/M system with the VPO, the data as file structure can be allocated on the MSU or the SSU for high speed access.

7.3 Effective Resource Management

The memory manager decides the number of the VP processes which use excessive amount of memory by limiting total of the storage size used by the VP processes. This size is decided by setting the highest address to a parameter called VPLIMIT. The VPLIMIT can be changed by the system administrator while the system is running.

7.4 High-Speed Swapping

The VPO can execute multiple VP processes up to the limit imposed by the VPLIMIT. The swapping function, which rolls out the lowest-priority process to the external storage units, enables the VPO to execute all the VP processes even if the total number of the processes is more than decided by the memory manager. In addition, the VPO transfers the data in parallel by multiple swapping devices to improve the performance. High speed swapping is also available between the MSU and the SSU to reduce the access time.

* UNIX is a registered trademark of AT & T.

8. LANGUAGE PROCESSING SYSTEM

The design goal of FORTRAN system for the VP system was to obtain high performance on the VP hardware as easy as on general purpose computers. FORTRAN77 EX/VP with new optimization features is developed on the basis of the current FORTRAN77/VP compiler.

8.1 Optimization

8.1.1 Parallel pipeline scheduling (PPS)

PPS facility reorders vector and scalar instructions to reduce idling time of the pipelines. Multiply & add operations are also recognized by this facility.

8.1.2 Loop unrolling

Loop unrolling is an optimization technique by which the iteration count of a loop is reduced and the operations within the loop are replicated for several subsequent index values. We added a new feature to unroll the innermost loops and to vectorize with regard to the outer loops. When the iteration count of the innermost loop is small, this loop is unrolled so that the operations in this loop can be vectorized as a part of an outer loop, resulting in a long vector length. Consequently, the vectorization ratio is increased and pipelines are used more efficiently. Figure 13 shows an example of matrix multiplication after vectorization and optimization.

8.2 Parallelization

FORTRAN77 EX/PP is a language processor system for parallel execution on QSP or DSP system of the VP2000 series hardware. There are two methods to develop parallel programs. The first method is the automatic parallelization for intra-procedural parallelism (DO loop slicing) and the second is the explicit parallelization using OCL (Optimization Control Line: compiler directive) for inter-procedural parallelism.

The automatic parallelization keeps portability of source programs to another system. Using the automatic vectorization with parallelization can derive high performance.

8.2.1 Automatic parallelization

Figure 14 shows examples of automatic DO loop slicing (parallelization). The standard facility of DO loop slicing is outer loop parallelization and inner loop vectorization (Fig. 14(a)). As extended feature, the FORTRAN77 EX/PP compiler selects an optimal DO loop for slicing among nested DO loops.

8.2.2 Parallelism description

a) Fork-join

Figure 15 shows an example of OCL to describe parallel execution of three subroutines. '!OCL PAR CALL' and '!OCL END PAR CALL' denote the starting and ending of parallel processing. They are the OCLs to execute the subroutines which are called between two OCLs concurrently. On the viewpoint of portability, these OCLs might be ignored and subroutines can be executed sequentially.

b) Synchronization

The FORTRAN77 EX/PP has three kinds of operation to synchronize subroutines executed in parallel. They are a mutual exclusion operation, a barrier operation and a post/wait operation.

The mutual exclusion is described by '!OCL MUTEX' and '!OCL END MUTEX'. The barrier and the post/wait are described by CALL statements for intrinsic subroutines.

9. PERFORMANCE

Table 3 shows the performance of famous benchmark program, LINPACK[8]. LINPACK is a program which is developed by Dr. J. Dongarra of Argonne National Laboratory of U.S.A. to compare the performance of different computer systems. This program can solve dense systems of linear equations using standard LU decomposition with partial pivoting and back substitution. The performance of LINPACK problem for a matrix of order 100 shown in Table 3 depends on the ability of the optimization features of The FORTRAN77 EX/VP compiler because the modification of the program is not allowed. To solve the problem for a matrix of order 1000, hand optimization of the program is allowed. On the VP2000 series, high performance is achieved by combination of hardware and software. According to the LINPACK report[8] dated 16th of February, 1991, these results are the best performance of the world attained by uni-processors.

10. CONCLUSION

This chapter described Fujitsu's new supercomputer, VP2000 series. The new features in the architecture, the Hardware implementation, the Multiprocessor system, the Hardware technology, the MSP system, the UNIX system, the Language processing system and the Performance were discussed. The detailed information concerning VP2000 series is available on the reports as refers as [5] and [6].

11. REFERENCES

[1] Miura, K. and Uchida, K., "FACOM VECTOR PROCESSOR VP-100/VP-200," High-Speed Computation, Kowalik, J.S. ed., Berlin/Heidelberg, Springer, 1984, pp. 127-38, (NATO ASI Series, Vol. F7).

[2] Uchida, K., Seta, Y., and Tanakura, Y., "The FACOM 230-75 Array Processor System," 3rd USA-JAPAN Comput. Conf., Oct., 1978, pp. 369-73.

[3] Uchida, K., Tamura, H., Takase, A. and Kuribayashi, N., "FACOM VP Series E model," ISR (Institute for Supercomputing Research) Technical Report, Proceedings of the first APPI work shop on supercomputing, 1987, pp. 202-220, (ISR TR87-03).

[4] Uchida, N., Hirai, M., Yoshida, M. and Hotta, K., " FUJITSU VP2000 Series," Compcon Spring '90, 1990, pp. 4-11.

[5] Uchida, N., Oinaga, Y., Tamura, H. and Shimizu, K., "System Overview of FUJITSU VP2000 series," Fujitsu Scientific and Technical Journal, Vol. 27, No. 2, 1991.

[6] Hotta, K., Kunai, T., and Honma, Y., "Basic Software for FUJITSU VP2000 Series", Fujitsu Scientific and Technical Journal, Vol. 27, No. 2, 1991.

[7] Miura, K., Nagakura, H. and Tamura, H., "VP2000 SERIES DUAL SCALAR AND QUADRUPLE SCALAR MODELS SUPERCOMPUTER SYSTEMS – A NEW CONCEPT IN VECTOR PROCESSING –," Compcon Spring '91, 1991.

[8] Dongarra, J., " Performance of Various Computers Using Standard Linear Equations Software," 1991(CS-89-85).

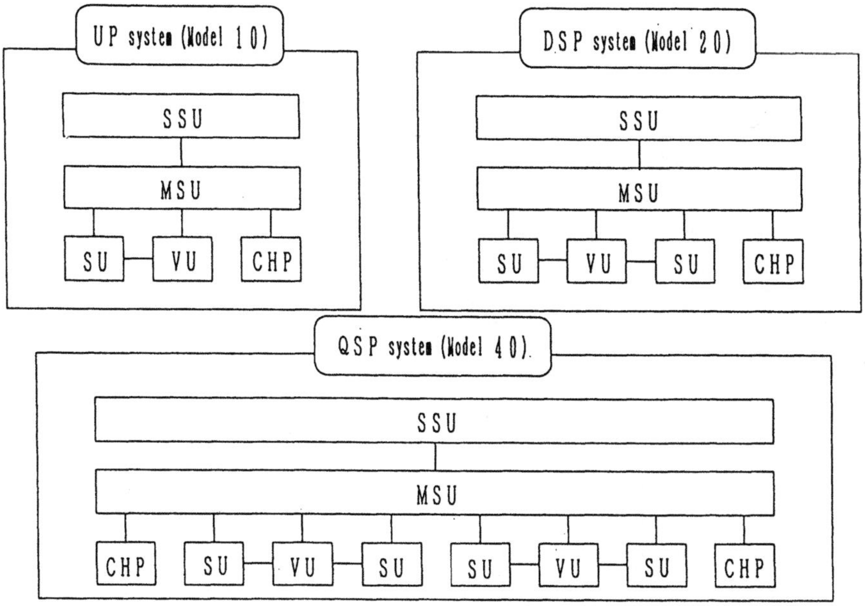

Fig. 1 System Configuration of Each Model

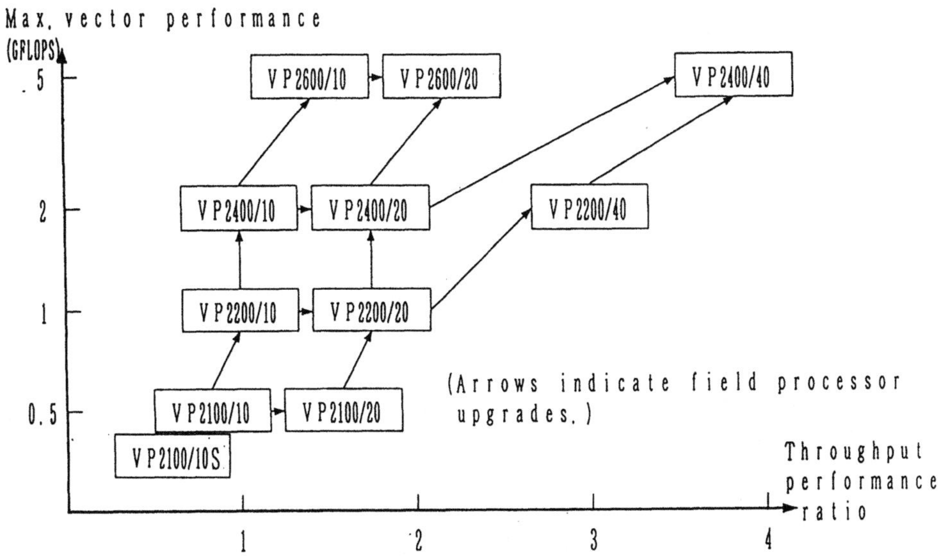

Fig. 2 Family of VP2000 Series

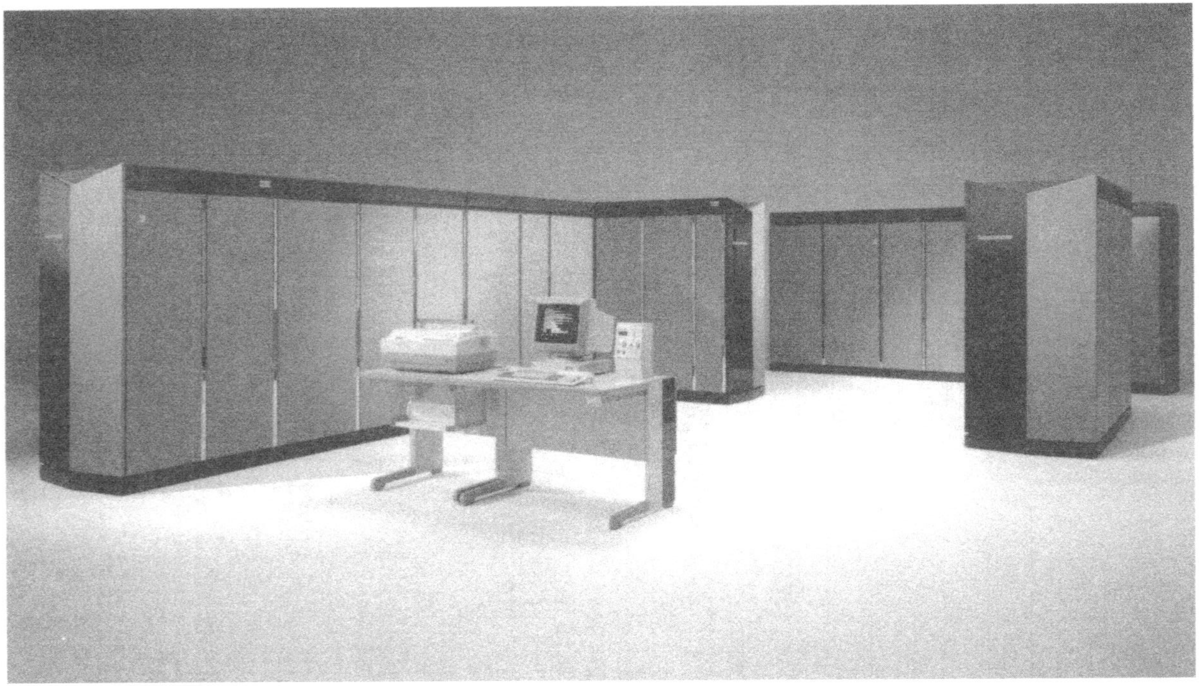

Fig. 3 Overall View of the VP2600/10

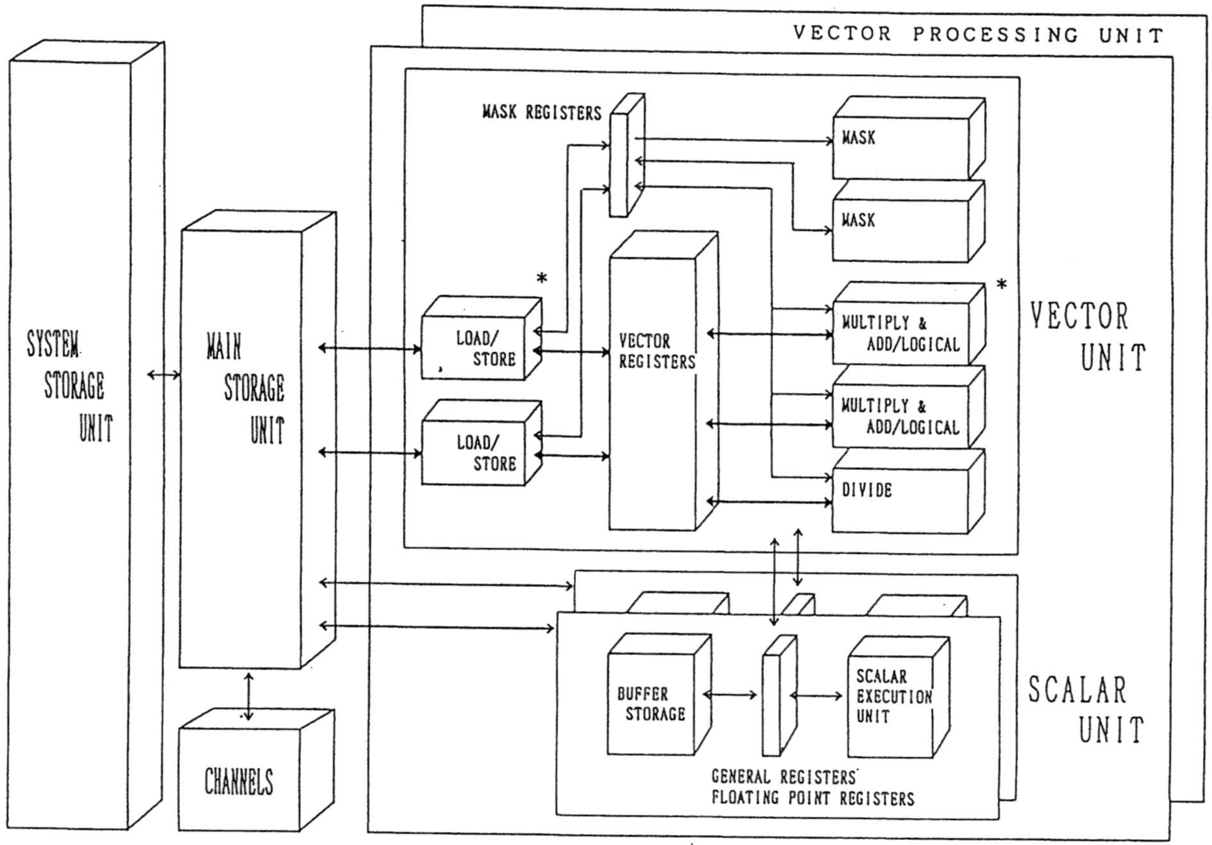

* : Not available for the VP2100 and VP2100S.

Fig. 4 VP2000 Series Hardware Block Diagram

V R format

0	8	16	24	32	40 47
OP	BOP	R2	R1	R3	R4

V X format

0	8	12	20	28	32	36 47
OP	BOP	R2	R1	X2	B2	D2

OP : Operation code (indicates VR format or VX format)

BOP: Extended operation code (indicates vector operation)

R : Register

X : Index register

B : Base register

D : Displacement

Subscript : Operand number

Fig. 5 Formats of Vector Instructions

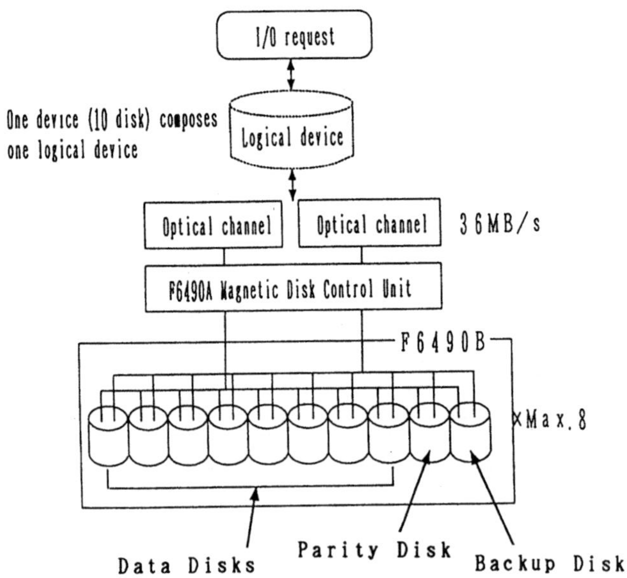

Fig. 6 Array Disk Subsystem

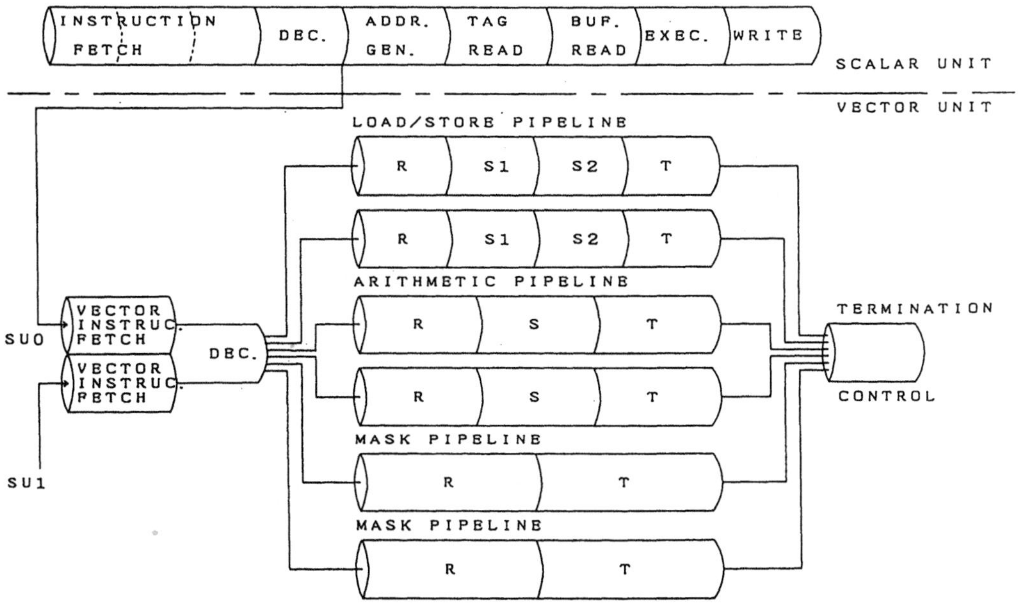

Fig. 7 Instruction Control Pipeline

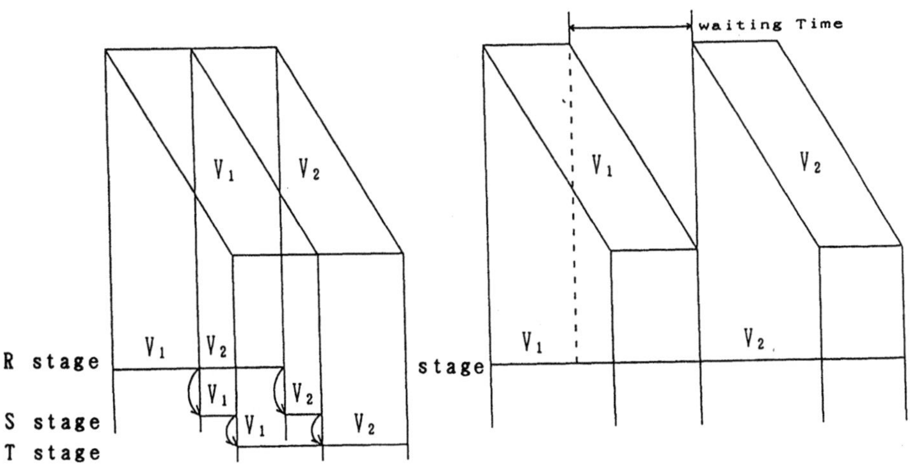

a) Continuous Processing b) Execution with Only One Stage

Fig. 8 Continuous Execution

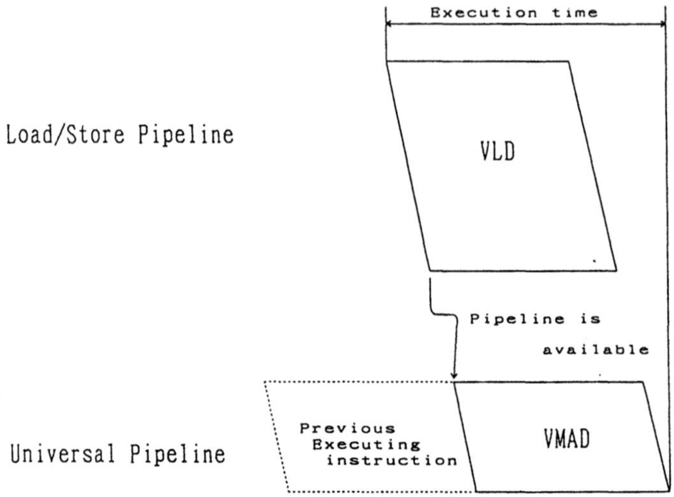

Fig. 9 Linkage Function

Fig. 10 15000 Gate-Logic LSI

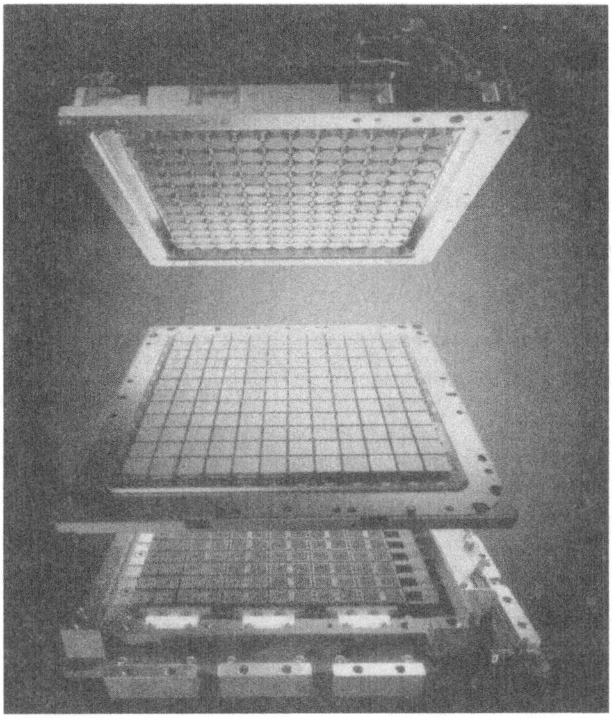

Fig. 11 Multi-Layer Glass Ceramics

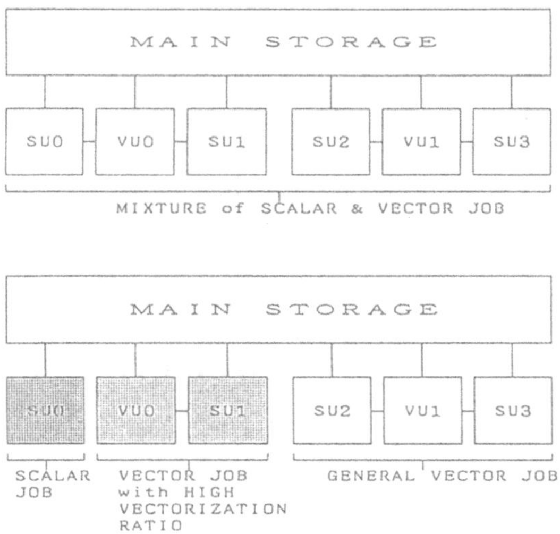

Fig. 12 Flexible System Configuration in QSP

〈Original program (matrix multiplication)〉

```
      DO 10 I=1,N
      DO 10 J=1,N
      DO 10 K=1,N
   10 A(I,J)=A(I,J)+B(I,K)*C(K,J)
```

⇩ Vectorization for nested DO loops·····Loop exchanging
⇩ Loop unrolling·························Unrolling for the secondary loop
⇩ Parallel pipeline sheduling···········Reordering instructions
Recognizing compound operations

〈Object code (pseudo instructions)〉

```
┌──→J
│     VLD     vr0,A(*,J)
│ ┌─→K
│ │   VLD     vr1,B(*,K)
│ │   VLD     vr2,B(*,K+1)
│ │   VMSAD   vr3,vr0,C(K,J),vr1
│ │   VMSAD   vr0,vr3,C(K+1,J),vr2
│ └──K=K+2
│     VSTD    vr0,A(*,J)
└────J=J+1
```

VLD : Vector load instruction
VMSAD: Compound inst. (Multiply and Add)
VSTD : Vector store instruction
vrn : vector register

〈Execution (Timing chart of vector pipelines)〉

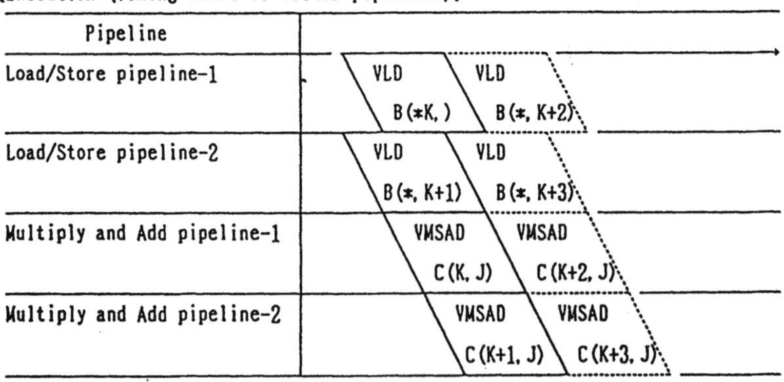

Fig. 13 An Example Vectorization and Optimization

```
P        DO 10 J=1,M
PV         DO 10 I=1,N
PV           A(I,J) = B(I,J) + C(I,J)
PV   10 CONTINUE
```

a) Inner Vectorization & Outer Parallelization

```
 V       DO 10 I=1,N
PV         DO 10 J=1,N
PV           A(I,J) = B(I,J) + C(I,J)
PV   10 CONTINUE
```

b) Optimal Index Selection

Fig. 14 DO Loop Slicing

```
!OCL  PAR CALL
      CALL SUB1(A)
      CALL SUB1(B)
      CALL SUB2(C)
!OCL  END PAR CALL
```

Make new instances of SUBROUTINE,
start parallel execution,
and wait ends of all instances.

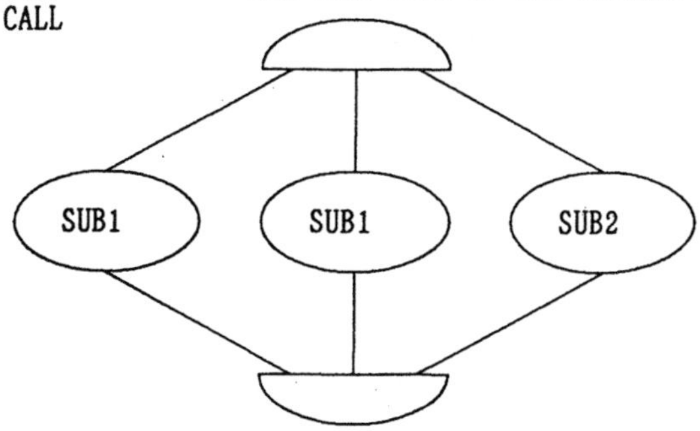

Fig. 15 Parallel CALL

Table 1 VP2000 Series Specifications

MODEL NAME		VP2100/10S	VP2100/20 VP2100/10	VP2200/20 VP2200/10	VP2400/20 VP2400/10	VP2600/20 VP2600/10	VP2200/40	VP2400/40
VECTOR PEAK PERFORMANCE(GFLOPS)		0.375	0.5	1	2	5	2	5
VPU	NUMBER OF VU	1					2	
	NUMBER OF SU	1	1~2				4	
MAIN STORAGE CAPACITY (MB)		32, 64, 128, 192, 256, 384, 512, 768, 1024		64, 128, 256, 384, 512, 768, 1024	128, 256, 512, 768, 1024	256, 512, 1024, 1536, 2048	128, 256, 512, 768, 1024, 1536, 2048	256, 512, 1024, 1536, 2048
SYSTEM STORAGE CAPACITY (GB)		1, 2, 4, 6, 8, 12, 16, 24, 32						
CHANNEL	NUMBER OF CHP	1					1~2	
	NUMBER OF CHANNELS	Max. 128					Max. 256	
	CHANNEL TYPE	MXC, BMC(4.5MB/s), OPTICAL CHANNEL(9MB/s), HIGH-SPEED OPTICAL CHANNEL(36MB/s), HIPPI CHANNEL(100MB/s)						
	THROUGHPUT	Max. 1GB/s					Max. 2GB/s	
VECTOR EXECUTION PIPELINE		M&A/L X1 DIVIDEX1		M&A/L X2 DIVIDEX1			M&A/L X4 DIVIDEX2	
LOAD/STORE PIPELINE		1		2			4	
VECTOR PIPELINE THROUGHPUT		1			2	4	1	2
VECTOR REGISTER CAPACITY		32KB/SU			64KB/SU	128KB/SU	32KB/SU	64KB/SU
BUFFER STORAGE CAPACITY		128KB/SU						

Table 2 Instruction Set

VECTOR INSTRUCTIONS		132
VXtype	LOAD/STORE	26
VRtype	ARITHMETIC OPERATION	36
	VECTOR RECURRENCE	3
	COMPARE	6
	LOGICAL OPERATION	13
	MACRO (Σ, FIND MAX, FIND INDEX etc.)	21
	VECTOR EDIT (COMPRESS, EXPAND etc.)	8
	VECTOR CONVERT	8
	CONTROL OPERATION	11

SCALAR INSTRUCTIONS	231

Table 3 Performance of LINPACK

	FUJITSU VP2600/10	FUJITSU VP2400/10	FUJITSU VP2200/10	FUJITSU VP2100/10
n=100	249	170	127	112
n=1000	4009	1668	842	445

unit=MFLOPS

III HITACHI S-820 SUPERCOMPUTER SYSTEM

Shun Kawabe
Kanagawa Works, Hitachi, Ltd.
1 Horiyamashita, Hadano, Kanagawa, 259-13 JAPAN

SUMMARY

The HITACHI S-820 made a debut as one of the most powerful supercomputers in the world in 1987. It consists of a Scalar Processor and a Vector Processor. As the Scalar Processor has the same architecture as a general-purpose mainframe, the whole complex fits well in conventional operating environments. Central to the high computation speed is the Vector Processor with its multiple-pipeline structure and large vector register memory. The use of the semiconductor Extended Storage dramatically reduces I/O time, contributing to faster job turn around and balanced system performance. A high degree of parallelism is incorporated inside the Vector Processor as well as between the Vector and the Scalar Processors. The hardware technology employed in the system, which is the key to high performance and supreme reliability, includes the field-proven state-of-the-art high-speed logic LSIs originally developed for Hitachi's large-scale mainframes, a 256 Kbits BiCMOS with an access time of 20 nsec, and a vector register LSI which combines logic and RAMs on a monolithic chip. A variety of software products has also been developed to fully exploit the hardware capabilities.

1. INTRODUCTION

As science and technology advance in scale and complexity, demand for sheer numerical calculation power is accelerating. Furthermore, as society becomes more sophisticated, the use of the supercomputer is spreading into a variety of fields thus far unthought-of. To meet increasing power demand and diversifying application needs, Hitachi has developed the S-820 supercomputer, with the design objectives:

(1) world-class performance,

(2) large system capacity,

(3) ease of use,

(4) high reliability,

(5) ease of installation.

Combining the state-of-the-art hardware technology with innovative logic design and the ad hoc vector architecture, the top-of-the-line model S-820/80 achieves a maximum arithmetic throughput of 3 GFLOPS (Giga FLoating-point Operations Per Second).

2. ARCHITECTURE AND SYSTEM ORGANIZATION

2.1 Overview

The S-820 has been designed to operate both in a loosely-coupled configuration and in a stand-alone configuration. In the former, the S-820 concentrates on numerical computation as a back-end subsystem, while the front-end mainframe executes conventional jobs such as compilation and performs system functions. In the latter, the stand-alone S-820 executes both front-end and back-end jobs by itself. Consequently, the S-820 processor complex comprises a scalar processor and a vector processor, along with other essential component units. As summarized in Fig. 1 the Scalar Processor employs the architecture of Hitachi's M-series mainframes in consideration of compatibility, connectivity and ease of migration. The Vector Processor, on the other hand, has its own vector architecture suited for high-speed engineering/scientific computation.

Scalar Processor		Vector Processor	
HITACHI M-series Architecture		Extended Vector Architecture	
M-series instruction set	216	Vector instruction set	90
General-purpose registers	16	Vector registers	32
Floating-point registers	16	Vector mask registers	16
Control registers	16	Scalar registers	32
Address translation		Vector address registers	48
Storage protection		Vector address translation	
TOD clock		Vector processing timer	
CPU timer			
Clock comparator			
Extended channel subsystem			

Fig. 1 Architecture of S-820

The vector architecture includes a set of 90 ad hoc instructions, a large stack of vector registers, a vector processing timer, and a large-capacity Extended Storage which serves as an internal storage for keeping intermediate results and for job swapping.

To provide budget-tight users with an easy entry to the supercomputer world and a smooth path of future growth, the S-820 is offered in five models: Model 15, 20, 40, 60 and 80 in the ascending order of performance. Table 1 presents the specifications of the S-820.

Table 1 S-820 Specifications

<table>
<tr><th colspan="2">Model</th><th>S-820/80</th><th>S-820/60</th><th>S-820/40</th><th>S-820/20</th><th>S-820/15</th></tr>
<tr><td rowspan="12">Scalar and Vector Processors</td><td>Peak Throughput of Arithmetic Units (GFLOPS)</td><td>3</td><td>1.5</td><td>0.75</td><td>0.375</td><td>0.25</td></tr>
<tr><td>No. of Vector Instructions</td><td colspan="5">90</td></tr>
<tr><td>General Registers</td><td colspan="5">16 (32bit)</td></tr>
<tr><td>Floating-point Registers</td><td colspan="5">16 (64bit)</td></tr>
<tr><td>Control Registers</td><td colspan="5">16 (32bit)</td></tr>
<tr><td>Vector Registers (words)</td><td>32*512</td><td>32*256</td><td colspan="3">32*128</td></tr>
<tr><td>Mask Registers (bits)</td><td>16*512</td><td>16*256</td><td colspan="3">16*128</td></tr>
<tr><td>Scalar Registers</td><td colspan="5">32 (64bit)</td></tr>
<tr><td>Vector Adress Registers</td><td colspan="5">48 (32bit)</td></tr>
<tr><td>Vector Processing Timer</td><td colspan="5">Yes</td></tr>
<tr><td>Buffer Storage (Kbytes)</td><td colspan="3">256</td><td colspan="2">128</td></tr>
<tr><td colspan="6"></td></tr>
<tr><td rowspan="6">No. of Vector Pipeline Elements</td><td>Add/Logical</td><td>4</td><td>2</td><td>1</td><td>1</td><td>1</td></tr>
<tr><td>Multiply/Add</td><td>4</td><td>2</td><td>1</td><td>1</td><td>1</td></tr>
<tr><td>Divide</td><td>1</td><td>1</td><td>1</td><td>1</td><td>1</td></tr>
<tr><td>Mask Operation</td><td>1</td><td>1</td><td>1</td><td>1</td><td>1</td></tr>
<tr><td>Vector Load</td><td>4</td><td>2</td><td>1</td><td>0</td><td>0</td></tr>
<tr><td>Vector Load/Store</td><td>4</td><td>2</td><td>1</td><td>1</td><td>1</td></tr>
<tr><td rowspan="2">Main Storage</td><td>Capacities (Mbytes)</td><td>128~1024</td><td>64~512</td><td>32~512</td><td>32~256</td><td>32~256</td></tr>
<tr><td>Error Checking/Correction</td><td colspan="5">1-bit error correction, 2-bit error detection</td></tr>
<tr><td rowspan="2">Extended Storage</td><td>Capacities (Gbytes)</td><td>0.5~24</td><td>0.5~12</td><td colspan="3">0.5~6</td></tr>
<tr><td>Error Checking/Correction</td><td colspan="5">2-bit error correction</td></tr>
<tr><td rowspan="2">Input/Output Processors</td><td>No. of Channels</td><td colspan="5">16~64</td></tr>
<tr><td>Max Transfer Rate</td><td colspan="5">288 Mbytes/sec</td></tr>
</table>

The Vector Processor can basically operate concurrently with the Scalar Processor. During compilation, a DO loop in the source code is converted (vectorized) to one or more vector instructions for the Vector Processor, and an instruction for starting the Vector Processor (EXVP(EXecute Vector Processing) instruction) is inserted into the code for the Scalar Processor. During the execution step, the Scalar Processor, upon encountering an EXVP instruction, activates the Vector Processor, which, in turn, executes the series of vector instructions associated with it, while the Scalar Processor proceeds with the subsequent scalar operation.

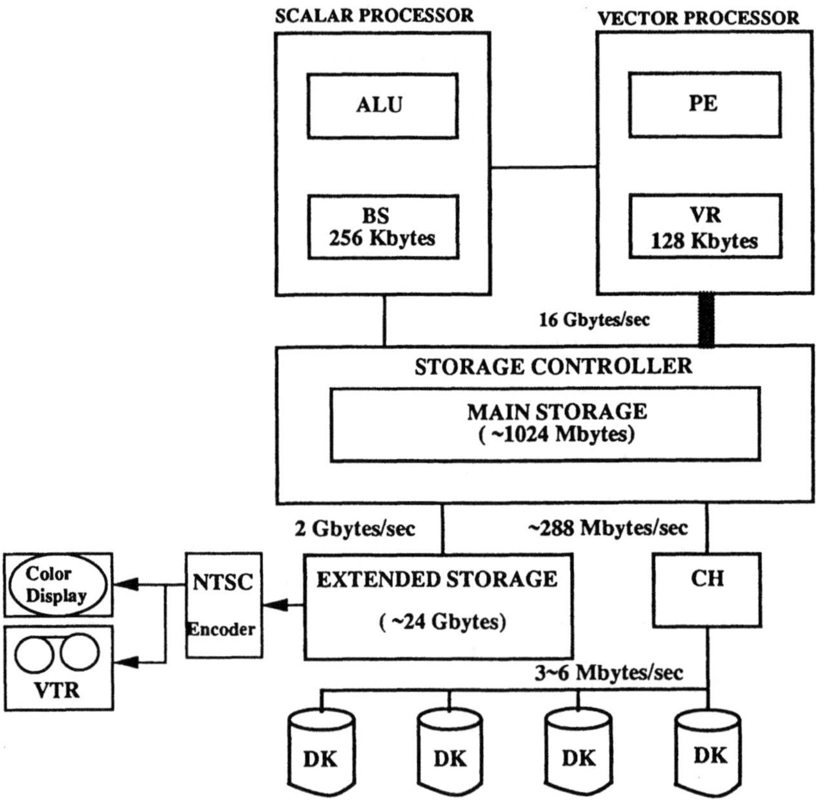

ALU : Arithmetic and Logical Unit
BS : Buffer Storage
CH : Channel
DK : Disk
PE : (Arithmetic) Pipeline Element
VR : Vector Registers

Fig. 2 Storage Hierarchy (S-820/80)

2.2 Extended Storage

Whereas the vector instruction set is one of the key architectural contributors to performance in terms of reduced CPU time, the Extended Storage dramatically reduces elapsed time, another popular measure of performance.

As illustrated in Fig. 2, the Extended Storage is hierarchically situated between the Main Storage, where programs and their data reside when they are to be executed, and the disk units, where they reside when they are not immediately needed. The Main Storage, made of fast semiconductor memory chips, tends to be limited in capacity due to cost impact. As a result, a program with a large amount of data does not fit at one time in the Main Storage, and its execution requires lengthy input/output (I/O) operations between the main storage and the disk(s) via a channel. For such large programs, the disk I/O limits the actual elapsed time, even if the vector performance is high.

The Extended Storage, also made of semiconductor memory chips, serves as a high-speed temporary storage for computation results. The capacity as large as 24 Gbytes often eliminates the need for disk I/O for many jobs, which results in significant reductions in elapsed time or job turn around time. Figure 3 well illustrates the effect of the Extended Storage: a five- to six-fold improvement in elapsed time performance has been achieved. The reduced frequency of running the I/O supervisor (part of the operating system) is also visible in the form of reduced CPU time.

Two synchronous instructions are provided for moving data (intermediate results) from the Main Storage to the Extended Storage and vice versa. "Synchronous" means that the Scalar Processor waits for the completion of this instruction, instead of proceeding further and being interrupted later. For the user's convenience, there is no need to rewrite the I/O statements in the FORTRAN source code: the compiler automatically (by the user's choice) replaces them with appropriate synchronous move instructions.

The Extended Storage also serves as a work storage for job swapping, which contributes to increase system throughput. Invisible to the user, the operating system is provided with an asynchronous move instruction, which is similar in format and operation to an input/output instruction.

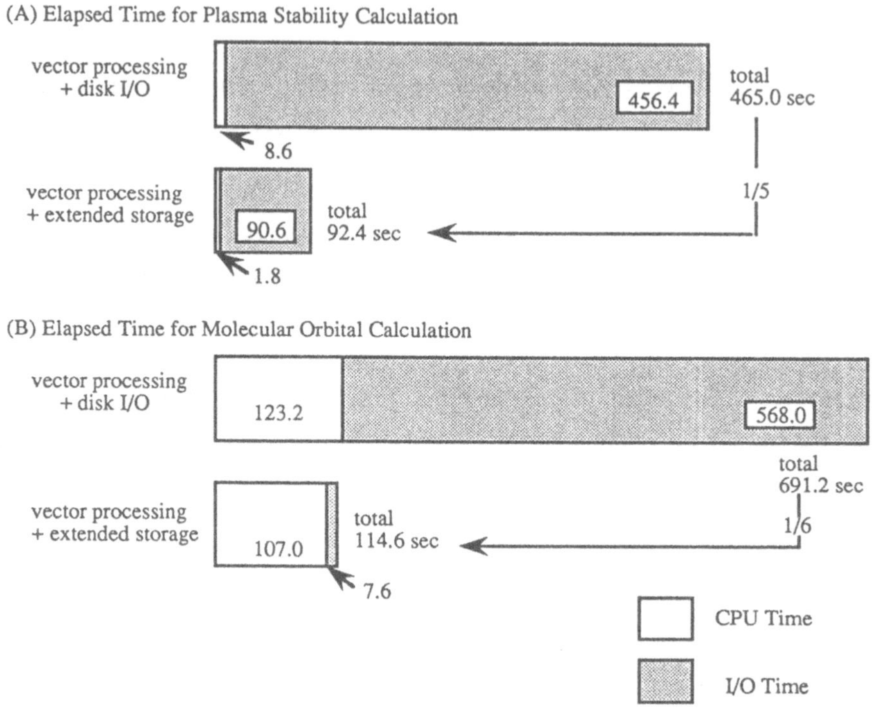

Fig. 3 Benefit of Extended Storage

2.3 Vector Register

During execution, vector operands reside in the Main Storage and are fetched into the Vector Processor as necessary. Likewise, the results of computation are stored into the Main Storage. In many instances, operands tend to be used more than once, and the result of computation tend to be referenced later in the same program. If operands and computation results are always to be stored in and fetched from the Main Storage, access contention would cause wait in the Vector Processor and impact performance, even with a large storage bandwidth.

Addressing this problem, the S-820 employs a vector register architecture: a large stack – as much as 128 Kbytes on Model 80 – of vector registers composed of high-speed bipolar memory LSIs is provided to hold a large amount of vector operands and intermediate results inside the Vector Processor which are likely to be re-used within a short period of time. As shown in Table 1, this is equivalent to 32 sets of vectors, each having a maximum of 512 64-bits elements.

2.4 Vector Instruction Set

To increase the vectorization ratio, i.e., to allow as many DO loops to be vectorized as possible, the S-820 has 90 specially defined instructions for vector operations. They include not only arithmetic instructions but also list vector instructions for gather/scatter operations, macroinstructions for compound operations, and compare instructions to generate mask vectors which allow vectorization of conditional (IF) statements. The Vector Processor contains 16 vector mask registers with up to 512 mask bits each (see Table 1).

3. LOGIC STRUCTURE

3.1 Overview

As illustrated in Fig. 4, the S-820 Model 80 processor complex consists of the following component units:

* one Scalar Processor,
* one Vector Processor,
* one Storage Controller,
* one Main Storage with the capacity ranging from 128 Mbytes to 1 Gbytes,
* one Extended Storage with the capacity ranging from 512 Mbytes to 24 Gbytes,
* one or two Input/Output Processors (IOPs), each of which contains from 8 to 32 channels,
* one Service Processor (SVP),
* one or two Console Devices.

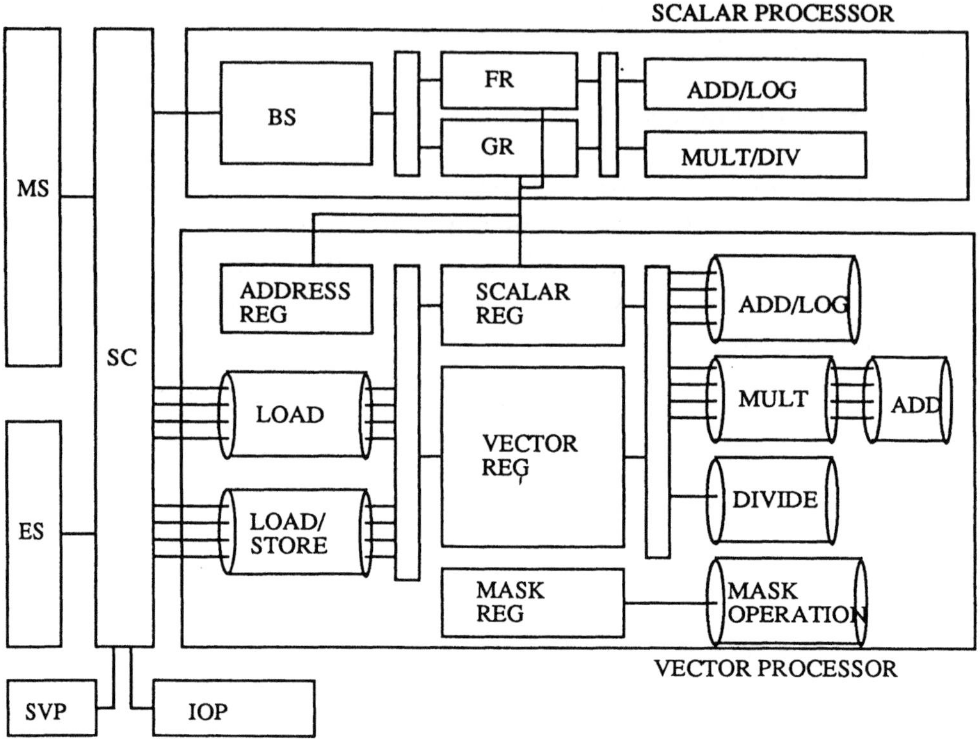

BS: Buffer Storage
ES: Extended Storage
FR: Floating-point Registers
GR: General Registers
IOP: I/O Processor
MS : Main Storage
SC : Storage Controller
SVP: Service Processor

Fig. 4 Processor Organization of S-820/80

The Scalar Processor is, in terms of logic structure, substantially similar to the Instruction Processor of Hitachi's large-scale mainframe M-680H.

The Vector Processor contains several floating-point arithmetic units (pipeline elements) and an array of vector registers, and executes vector instructions at high-speed; its structure and operation will be discussed in more detail in Section 3.2. In Fig. 4, the four lines coming in or out of each resource (denoted by a cylinder) means that there are four identical units each for Model 80.

The Storage Controller performs data transfer between the Main Storage and any of the Extended Storage, the Scalar Processor, the Vector Processor and IOPs, solving contentions among them. The Main Storage serves as an ordinary processor storage which contains programs, data and system tables. The Extended Storage constitutes an intermediate level in the storage hierarchy and provides excellent potentials for significant elapsed time and system performance improvements.

The channels, housed in Input/Output Processors (IOPs), transfer data between the Main Storage (via the Storage Controller) and input/output devices, over an industry-standard protocol. The microprocessor-based Service Processor (SVP), in combination with a Console Device (CD) attached to it, monitors and controls the operation of the processor complex and performs system functions such as Initial Program Load (IPL), and Reliability, Availability and Serviceability (RAS) functions.

3.2 Vector Execution Control

This section discuss the design features of the Vector Processor from the performance point of view.

3.2.1 Parallel construction

As illustrated in Fig. 4, the S-820 Model 80 has four add/logical operation units, four multiply-and-add units and a divide unit for arithmetic operations. A multiply-and-add unit consists of a multiply subunit and an add subunit in a cascade connection. The twelve (12) arithmetic units, namely, the add/logical operation units, the multiply subunits and the add subunits, together with the load or the load/store units and rows of vector registers, form pipelines in that operands go through them vector element after vector element in a pipeline manner. Each of these units/subunits is called a pipeline element. The minor models (Models 60, 40, 20 and 15) have fewer pipeline elements.

The pipeline elements can operate simultaneously with each other. Each pipeline element produces a 64-bits result every four (4) nanoseconds (nsec), which is the clock cycle of the Vector Processor. The S-820 Model 80 has altogether 12 pipeline elements, and hence has a maximum arithmetic throughput of 3 GFLOPS.

The maximum number of arithmetic operations performed per instruction per machine cycle on a pair of operands (vector elements) by the S-820's vector instruction set is two (2). Therefore, the maximum single-instruction execution speed is 2 GFLOPS, as in the case of the Vector Multiply and Add instruction. However, thanks to the parallel structure of the add/logical units and the multiply-and-add units, two dissimilar instructions can be executed simultaneously, one using the former and the other using the latter, in which case the maximum arithmetic throughput of 3 GFLOPS can be achieved. Although a carefully coded synthetic program has scored more than 2.5 GFLOPS as sustained performance, real-life programs will see somewhat lower performance due to various overhead and the difficulty in taking full advantage of the parallelism all the time. The divide unit cannot operate simultaneously with other arithmetic units/subunits.

Table 2 shows arithmetic execution speeds of some vector instructions on the S-820 Model 80.

Table 2 Vector Instruction Execution Speed (S-820/80)
(on 64-bits floating-point operands)

unit: MFLOPS

Add	1000
Multiply	1000
Divide	250
Multiply and Add	2000

3.2.2 Elementwise parallel processing

One of the load units, one of the load/store units, a pair of rows of vector registers and one of the arithmetic units form a pipeline, also called a section; there are multiple pipelines in Models 80 and 60.

One way to take advantage of the parallel pipelines is to execute multiple vector instructions on them. This is not always easy, however, since complete mutual independency of these instructions must be guaranteed.

The S-820 employs elementwise parallel processing, in which one vector instruction monopolizes all the pipelines of one kind (add/logical or multiply-and-add), allocating different vector elements to different pipelines. Figure 5 illustrates elementwise parallel processing on Model 80 with four pipelines (sections), taking a vector inner (scalar) product operation between two 80-element arrays (operands) as an example.

$$S = A1*B1 + A2*B2 + \cdots\cdots + A80*B80$$

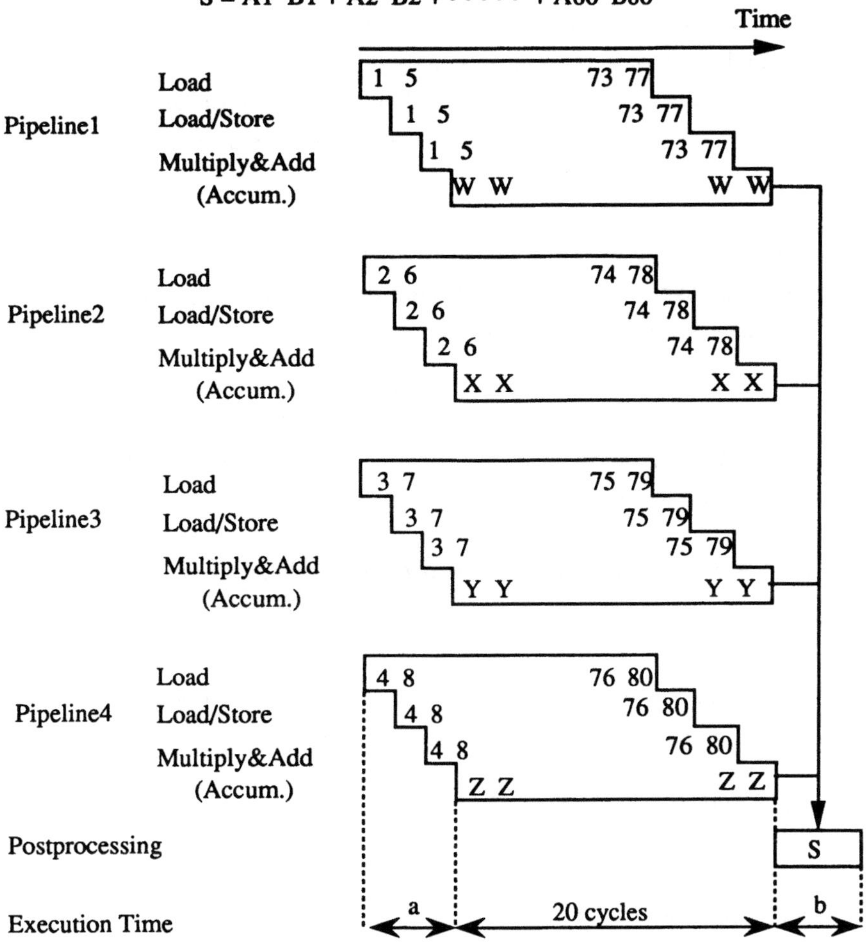

Accum.: Accumulator inside Multiply-and-Add Unit

W = A1*B1 + A5*B5 + • • • • • • • + A77*B77
X = A2*B2 + A6*B6 + • • • • • • • + A78*B78
Y = A3*B3 + A7*B7 + • • • • • • • + A79*B79
Z = A4*B4 + A8*B8 + • • • • • • • + A80*B80
S = W + X + Y + Z

Fig. 5 Flow of Elementwise Pipeline Processing

The vector elements of Array A (first operand) are loaded into four rows of vector registers via four load units; the vector elements of Array B (second operand) are loaded into another set of four rows of vector registers via four load/store units; multiplication and addition are performed on four pairs of vector elements simultaneously in the four multiply-and-add units. In each pipeline, the result of multiplication is accumulated into the accumulator inside the multiply-and-add unit one cycle after another. When all the vector elements (operands) are exhausted, the four partial sums contained in the four accumulators are added together to produce the final result S at the post-processing stage.

As indicated, the total execution time is composed of the start up time a, the core execution time (20 cycles), and postprocessing time b. Generally, with a vector size of N, the total execution time Te is expressed as follows:

$$Te = a + (N/4) + b.$$

a and b do not depend upon the vector size, i.e., the number of vector elements per operand (array), and are relatively small.

Without elementwise parallel processing, the total execution time Tn would be:

$$Tn = a + N + b.$$

With a large size of the vector (array), $Te \approx Tn/4$, i.e., elementwise parallel execution is nearly four times as fast as serial pipeline execution.

3.3 Storage Control

Generally, the structure of a system's internal storage must be so designed as to provide a sufficient bandwidth, i.e., must be able to feed the arithmetic units with data as fast as required by them. Otherwise, the arithmetic units would wait for the arrival of data, and this loss time would surface as performance overhead.

The bandwidth between the Storage Controller (Main Storage) and the vector register stack of the S-820/80 is 16 Gbytes/sec, which may appear to be rather small when one considers the maximum arithmetic throughput of 3 GFLOPS: on the average 2*2 (operands) *8 (bytes) = 32 Gbytes/sec might be "sufficient". (Note that each multiply-and-add pipeline requires only one – instead of two – pair of vector operands for two operations in a machine cycle. Thus, 2*2*8 instead of 3*2*8. Also, when one is to consider result storing as well, this should be a little bit higher.) Most actual programs, however, do not exercise the arithmetic units at maximum duty all the time. They contain some scalar operands as well, which do not require Vector Processor -to-Main Storage accesses. The existence of the vector registers helps significantly reduce storage access frequency since operands and intermediate results tend to be re-used within the same program. On the whole, the bandwidth of 16 Gbytes/sec should be sufficient for the S-820/80 in practical sense.

Even with a sufficient bandwidth, some loss of time may be observed from time to time due to contention at resources, especially access contention at the Main Storage, when a particular storage bank is congested with multiple access requests. Such contention also shows up as performance overhead, which means reduced vector performance.

To reduce as much access contention as possible, the S-820 organizes the banks of the Main Storage with a carefully designed addressing scheme in consideration of a cyclic nature of the operand address pattern.

4. HARDWARE TECHNOLOGY

Whereas the number of machine cycles per instruction (operation) can be generally reduced by increasing the amount of logic circuits and improving the control structure, hardware technology is the key factor determining the bare hardware speed, which is often represented by machine cycle.

In addition to the state-of-the-art logic and memory LSIs used in Hitachi's large-scale mainframe M-680H, the S-820 employs several newly developed semiconductor components. Essential in achieving the machine cycle time of 4 nsec is the VR (Vector Register) LSI, which is built on bipolar technology and contains 6912 bits of memory and 2500 gates of logic including the data registers. The chip access time is 2.5 nsec to allow a vector register access pitch (cycle) of 4 nsec required by the logic (arithmetic unit).

A 64 Kbits and a 256 Kbits BiCMOS memory chips have been developed for the Main Storage, with a chip access time of 20 nsec, to satisfy the required bandwidth. A 256 Kbits and a 1 Mbits CMOS memory chip with a chip access time of 45 nsec are employed in the Main Storage of the economical version of the S-820. The Extended

Storage uses industry-standard 1 Mbits and 4 Mbits CMOS chips.

To allow cooling only by air, the S-820 packages an LSI chip on a ceramic substrate with high thermal conductivity; these substrates are then surface-mounted on a multi-layer printed circuit board called a package.

Table 3 summarizes the hardware technology used in the S-820.

Table 3 Hardware Technology Used in S-820

Component	Characteristics
LSI (bipolar)	2000 / 5000 gates 0.2 / 0.25 nsec
RAM (bipolar)	4 K / 16 Kbits 4.5 / 12 nsec
VR LSI	2500 gates + 6912 bits 2.5 nsec
MS SRAM (CMOS) (BiCMOS)	256 K / 1 Mbits 45 nsec 64 K / 256 Kbits 20 nsec
ES DRAM	1 M / 4 Mbits 120 nsec
PK (package)	72 LSIs / PK 22 layers
PL (platter)	21 PKs / PL 22 layers

Table 4 Major Products Available on S-820

Category	Name	Description
System Software	VOS3/HAP/ES	System control program based on general-purpose mainframe operating system
	FORT77/HAP	Vectorizing FORTRAN compiler supporting FORTRAN77 language specification
	FORT/ASSIST FORT/VF FORT77/TD	Tools for programming, debugging and tuning
Application Software	DEQSOL E2	Differential equation solver based on a high level language
	SGRAF	Program which visualizes simulation results
	MATRIX/HAP	Library of vectorized matrix calculation subroutines
	ISAS II	Structural analysis package

5. SOFTWARE

A powerful set of easy-to-use software packages have been developed to help the user take full advantage of the S-820's hardware capabilities. Table 4 lists major program products available on the S-820.

VOS3/HAP/ES, the primary control program, is a multiple virtual space operating system which supports the vector architecture in addition to the Hitachi M-series (scalar) architecture. It drives vector jobs in parallel with conventional scalar jobs, both in a stand-alone and in a loosely-coupled configuration, while managing the Main Storage and the Extended Storage efficiently.

FORT77/HAP compiles (vectorizes) FORTRAN source code into object code with high efficiency, as discussed in Section 6.2 in detail.

DEQSOL E2 directly accepts differential equations in their original mathematical expressions, dramatically reducing programming chores. With this product, the number of lines (statements) in source code can be an order of magnitude smaller, and the program becomes very visible and manageable.

In addition, a variety of off-the-shelf, third-party application packages are being converted to vector versions for the S-820.

A compiler with a powerful vectorizing capability is indispensable in exploiting hardware performance potentials of any supercomputer. The vectorizing FORTRAN compiler FORT77/HAP makes full use of the advanced architecture of the S-820 and employs various techniques in achieving the performance objective.

To increase vectorization ratio, which is the key indicator of object code efficiency, FORT77/HAP first takes advantage of the rich vector instruction set:

* Use a compare instruction, which generates a mask vector for elementwise selection of operation/no-operation, in conjunction with a conditional ("IF") statement.
* Use a list vector instruction for indirect addressing in which the subscript of a vector(array) is another vector (array).
* Use a vector macroinstruction for a compound operation consisting of multiple operations, such as sum of products, inner (scalar) product, first order iteration and element sum.

To further improve vectorization ratio, it then attempts to minimize and isolate the unvectorizable portions by rearranging source statements. For a DO loop which it failed to vectorize because it could not resolve data dependencies, it also analyzes the flow of the data outside the DO loop to determine whether the DO loop is inherently unvectorizable. Take the following program, for example:

```
            K1=K+1                        (1)
              .
              .
            DO   10   I=1,N
              A(I+K1)=  .  .  .  .  .     (2)
                      =A(I+K)             (3)
10          CONTINUE
```

Looking only at statements (2) and (3), it is impossible to determine which subscript, I+K1 or I+K, is larger than the other; therefore this loop cannot be vectorized. By expanding the scope of analysis beyond the boundaries of the DO loop and examining statement (1), the compiler can now determine that K1 > K, and can vectorize the DO loop.

FORT77/HAP attempts to vectorize not only unit DO loops but also multi-fold DO loops. Here, a stand alone (single-fold) DO loop and the innermost one in a nest of DO loops (multi-fold DO loop) are called a unit DO loop. The above-discussed techniques apply to unit DO loops. FORT77/HAP uses the following techniques to vectorize multifold DO loops:

* Exchange the innermost loop with the outer loop.
* Decompose the multi-fold loop into a unit loop.
* Divide the multi-fold loop.

The following example illustrates the exchanging of the innermost loop with the outer loop:

before

```
          DO 20 J = 1,N
          DO 10 I = 1,N
            A(I,J) = A(I-1,J) + A(I-2,J)
10        CONTINUE
20        CONTINUE
```

This loop cannot be vectorized because one statement contains I, I-1 and I-2 as subscripts (indexes), with I being varied.

after

```
         DO 20 I = 1,N
         DO 10 J = 1,N
           A(I,J) = A(I-1,J) + A(I-2,J)
10       CONTINUE
20       CONTINUE
```

This new innermost loop can be vectorized.

Another example shown below, which applies the Gaussian elimination method to a system of linear equations, demonstrates how dividing a multi-fold DO loop can make it vectorizable.

before (cannot be vectorized)

```
         DO 200 J = K + 1, N
              IF (IPK .NE. K) THEN
              W = A(IPK,J)
              A(IPK,J) = A(K,J)
              A(K,J) = W
              END IF
         DO 100 I = K + 1, N
              A(I,J) = A(I,J) + A(I,K) * A(K,J)
100      CONTINUE
200      CONTINUE
```

after (can be vectorized)

```
         DO 201 J = K + 1, N
              IF (IPK .NE. K) THEN
              W = A(IPK,J)
              A(IPK,J) = A(K,J)
              A(K,J) = W
              END IF
201      CONTINUE
         DO 110 J = K + 1,N,2
         DO 100 I = K + 1, N
              A(I,J) = A(I,J) + A(I,K) * A(K,J)
              A(I,J + 1) = A(I,J + 1) + A(I,K) * A(K,J + 1)
100      CONTINUE
110      CONTINUE
```

Incidentally, this last example uses also the technique of unrolling, which expands one statement into multiple statements in order to minimize software overhead associated with loop control for the "DO 200" loop by reducing the loop count. Basically, the scalar or unvectorized loop, in this case the "DO 110" loop, benefits from unrolling, as the S-820 employs elementwise parallel processing and expanding one statement into multiple statements does not particularly increase parallelism in the Vector Processor.

6. PERFORMANCE

Table 5 shows the scalar and vector performance of the S-820/80 for basic operations. The vector performance is indicated as a function of vector length. At vector length 1000, the vector performance is from 5 to 70 times as high as the scalar performance. The S-820 demonstrates excellent vector performance, because the effect of start up time required for loading vector operands becomes relatively smaller and elementwise parallel processing becomes more effective. It is worth noting that the S-820 achieves 80 % of the peak performance (3 GFLOPS) at a vector length of 1000 for polynomial operations, which make full use of multiple arithmetic units. It should be noted that the performance for list vector operations is as good as that for simple addition or multiplication.

Table 5 Measured MFLOPS of basic operations

Operation	Fortran	Scalar	Vector length			
			50	100	300	1000
Add	A(I)=B(I)+C(I)	13.3	49.8	92.3	221.5	411.4
Multiply	A(I)=B(I)*C(I)	12.6	49.1	98.1	232.3	400.1
Divide	A(I)=B(I)/C(I)	4.7	38.5	67.9	130.9	184.6
Inner Product	SUM=SUM+B(I)*C(I)	19.3	59.6	114.3	300.1	686.1
Multiply and Add	A(I)=A(I)+B(I)*C(I)	19.6	90.9	169.4	384.1	669.7
Polynomial operation	V=V+S*V+S*V+S (*5 lines)	62.1	960.1	1488.3	2149.3	2410.2
Polynomial operation	SUM=SUM+(B(I)+C(I))**2 (*4 lines)	26.5	303.2	533.2	1252.3	1919.7
IF-statement	IF ( ) THEN A(I)=A(I)+B(I)*C(I) ELSE A(I)=A(I)-B(I)*C(I)	8.5	68.6	126.3	242.1	351.4
List vector	A(I)=B(I)+C(J(I)+K)*D(I)	8.5	100.1	181.1	384.1	400.1
1st order iteration	A(I)=A(I-1)-B(I)*C(I)	20.4	41.7	62.5	90.3	106.3

7. CONCLUSION

The HITACHI S-820 has been developed as one of the most powerful supercomputers in the world, delivering a peak arithmetic throughput of 3 GFLOPS on a single vector processor. The world class performance has been achieved by first analyzing the factors determining performance and then effecting a number of improvements in architecture, logic design, compiler and hardware technology. The key performance features include the advanced vector instruction set, the Extended Storage, elementwise parallel execution and the intelligent vectorizing compiler FORT77/HAP.

8. REFERENCES

[1] Kawabe, S., "HITACHI Supercomputer S-820 System Overview," Proceedings of the First Appi Workshop on Supercomputing, 221-230, 1987.

[2] Wada, H., Kawabe, S. and Odaka, T., "HITACHI Supercomputer S-820 Overview," Supercomputing Europe 89, 139-147, 1989.

[3] Odaka, T. et al., "Development of Hitachi Supercomputer S-820 System," Proceedings of the Third International Conference on Supercomputing, Boston, 71-77, 1988.

[4] Wada, H. et al., "High-speed Vector Instruction Execution Schemes of Hitachi Supercomputer S-820 System," Proceedings of the International Conference on Parallel Processing, 291-298, 1988.

[5] Wada, H. et al., "Performance Enhancement by Extended Storage on Hitachi Supercomputer S-820 Systems," Proceedings of the Fourth International Conference on Supercomputing, Santa Clara, 1989.

[6] Eoyang, C. et al., "The Birth of the Second Generation: The Hitachi S-820/80," Proceedings of Supercomputing '88, 296-303, 1988.

IV NEC SX-3 SUPERCOMPUTER SYSTEM

Tadashi Watanabe
NEC Corporation
5-7-1, Shiba, Minato-ku, Tokyo, 108-01 JAPAN

1. INTRODUCTION

The NEC SX-3 supercomputer system is a high-speed and large-scale supercomputer system employed multiple vector pipelines and multiprocessor configurations. The system realized a machine cycle time of 2.9 nsec by employing high-speed 20000 gate VLSI chip with 70 psec switching speed and high-density packaging system. The SX-3 supports the Super-UX operating system based on the AT & T System V UNIX and the vectorizing and parallelizing Fortran compiler.

The SX-3 system consists of eight models. The models are based on the number of processors, and the number of vector pipelines installed per processor. The models range from a 690 MFLOPS uniprocessor entry model to the 22 GFLOPS quad processor top end model. Table 1 summarizes SX-3 specifications. This chapter introduces the system configuration, processor architecture, features of Fortran and tools, and recent results of benchmarking tests.

Table 1 Specification of SX-3

MODEL	SX-3								SX-2A
	44	42	24	22	14	12	11	1L	
Vector Peak Performance	22GFLOPS	11GFLOPS		5.5GFLOPS		2.75 GFLOPS	1.37 GFLOPS	0.68 GFLOPS	1.3 GFLOPS
Number of Arithmetic Processors	4		2		1				1
Vector Registers	144KB	72KB	144KB	72KB	144KB	72KB	36KB	36KB	80KB
Vector Mask Registers	256b x 8	128b x 8	256b x 8	128b x 8	256b x 8	128b x 8	64b x 8	64b x 8	256b x 8
Scalar Registers	64b x 128								
Vector Pipelines	16	8	16	8	16	8	4	2	16
Cache Memory	64KB								
Main Memory [Max]	2048MB			1024MB			512MB		1024MB
Extended Memory	16GB							8GB	8GB
I/O Channels [Max]	256							64	64
I/O Transfer Rate	1GB/sec							192MB/s	192MB/s

2. SYSTEM CONFIGURATION

Figure 1 shows a maximum configuration of the SX-3 system. The Arithmetic Processor (AP) is a main processor which has a powerful vector and scalar unit for executing user codes. The system can configure four arithmetic processors at the maximum configuration sharing the main memory unit with a maximum capacity of 2 Gbytes. The control processor handles I/O buffering and controls. The Input/Output Processor (IOP) has up to 64 channels each through which various peripherals such as disk units and laser printers are connected. In addition, up to eight pairs of HIPPI (High Performance Parallel Interface) channels with a transfer rate of 100 Mbytes/sec each are supported through HIPPI adapter. The Extended Memory Unit (XMU) is a large capacity semiconductor storage unit of up to 16 Gbytes with a transfer speed of 2.75 Gbytes/sec which functions as data files, swapping files and disk cache.

3. PROCESSOR CONFIGURATION AND ARCHITECTURE

Figure 2 shows the internal structure of the Arithmetic Processor (AP). The AP is functionally divided into two units, that is, the scalar unit and vector unit.

The major parts of the scalar unit are 64 Kbytes cache, 4 Kbytes instruction buffer, 128 scalar registers, and one set of scalar pipelines. For achieving fast scalar processing, the system employs RISC (Reduced Instruction Set Computer) architecture based on the simplified hardware design and control.

The cache stores instructions and scalar data. The instruction buffer with 4 Kbytes storage provides high-speed access to instructions, particularly instructions in DO loops can be usually fetched from this instruction buffer without extra access to the cache or main memory. Corresponding to each entry of the instruction buffer, there exists an entry of a branch history buffer that memorizes a branch direction of the previously executed branch instruction. According to the memorized direction (go/no go), the processor control hardware prefetches the instruction and processes it in advance. That is called a branch prediction mechanism.

Normally, the instructions are served based on 'first come, first serve', but when some instruction (A) happened to have resource conflict with the previous instruction and the next instruction (B) has no resource conflict, then the next instruction (B) is controlled to execute before the execution of the instruction (A). This is just like a data flow control, i.e., hardware controlled instruction reordering. This feature greatly reduces the loss due to waiting time for resources. Figure 3 shows the effects of scalar pipeline and instruction reordering.

The vector unit consists of four identical sets of vector pipelines at the maximum configuration. Each set has two add/shift pipelines and two multiply/logical pipelines clocked in 2.9 nsec. There are 16 vector pipelines in total. The vector operations are handled in an interleaved way across four sets of vector pipelines, but four functional pipelines in each set work independently, that is, execute four different vector instructions. When a DO loop includes two different vector adds like $A(I) = B(I) + C(I)$ and $D(I) = E(I) + F(I)$, those two vector adds are executed in parallel utilizing two add pipelines in each set as shown in Fig. 4. To flow vector data in and out smoothly and quickly, the processor has two vector loads and one vector store paths, each of which has a maximum of 4 words (256 bits) width and works independently.

There are 72 vector registers that are divided into two parts, Vector Arithmetic Register (VAR) and Vector Data Register (VDR). The VAR consists of 256 words $\times$ 8 registers for vector arithmetic operations. The VDR works as high-speed programmable vector cache that is reconfigurable with three configurations of 256 words $\times$ 64, 128 words $\times$ 128 or 64 words $\times$ 256 according to the DO loop characteristics. Those configurations are dynamically changeable during the program execution without performance loss.

4. THE SUPER-UX OPERATING SYSTEM

The Super-UX operating system was developed based on AT & T System V with many extensions and enhancements to meet requirements for supercomputer system as shown in Fig. 5. Major extensions are:

* Multiprocessor control function,
* Support of large and high-speed I/O devices such as extended memory unit,
* High-speed I/O functions including bufferless I/O, parallel I/O and disk cache,
* Large-scale file system with virtual volume functions,
* System operation management such as accounting, file space management and

file back-up function.

5. FORTRAN AND TOOLS

The Fortran supported in SX-3 has auto-vectorizing and parallelizing functions. Table 2 shows typical vectorizing features of the Fortran. To make vectorization easier, various vector instructions are supported such as gather/scatter, compress/expand, summation, find max/min, and masked vector operations. Figure 6 shows several examples of vectorized loops that are taken from LINPACK and Livermore Fortran Kernels (LFK). In this figure, "V", "X", "W" and "C" on the left side show that those loops are vectorized, interchanged and vectorized, collapsed and vectorized, and conditionally vectorized, respectively.

Table 2 Automatic Vectorizing Features

BASIC VECTORIZATION		
HIGHLIGHTS	Loops Including	
	IF Statements	
	Vector Subscripts	A(IX(I))=B(IX(I))+C(IX(I))
	Macro Operations	
	Summation	S=S+A(I)
	Inner Product	S=S+A(I)*B(I)
	Product	S=S*A(I)
	Iteration	A(I+1)=X*A(I)+B(I)
	Find Max/Min	Finding Max/Min Value and/or Its Index
	Search	Table Lookup
	Compress/Expand	Compression/Expansion of Matrix

EXTENDED VECTORIZATION			
FOR HIGHER VECTOR RATIO	Loops Including		
	Unvectorizable Statements	→	Partial Vectorization
	Data Dependencies	→	Statement Interchange Introduction of Work Vectors
	Subprogram Calls	→	Inline Expansion
	Unknown Dependencies	→	Conditional Vectorization
	Nested Loops	→	Vectorization of Outer Loops
FOR EFFICIENT VECTORIZATION	Short Loops with Constant Iteration	→	Loop Expansion
	Small Loops with Few Operations	→	Loop Unrolling
	Tightly-nested Loops	→	Loop Collapsing Loop Interchange
	Loops with Variable Length	→	Conditional Vectorization

In addition, various optimization functions are supported for full hardware utilization. They include the in-line expansion of subroutine, loop unrolling and instruction reordering.

As for the parallel processing or multitasking, two types of functions are provided; macrotasking and microtasking as shown in Fig. 7. The macrotasking is a subroutine level multitasking supported by a collection of Fortran callable library routines shown in Table 3. In this table, PTFORK and PTJOIN are used for task control by invoking a subtask and waiting for the completion of subtasks. PLLOCK and PLUNLOCK are routines for exclusive control and task synchronization (See Fig. 8). The macrotasking is a top-down task partitioning for parallel processing where a global level of parallelism is needed, and the restructuring of algorithm may be sometimes required. On the other hand, the microtasking is a loop or statement level multitasking realized by mainly automatic parallelization and the parallelizing directives. Table 4 shows some examples of parallelization primitives for microtasking. In this table, "Parallel DO" is used for DO loop parallelization when the compiler cannot find the parallelizable part automatically. "Parallel Case" is for the statement level parallelization. "Critical Section", "Serial Section" and "Barrier" are primitives for synchronization and exclusive controls.

Table 3 Macrotasking Library Routines

GROUP	FORM	FUNCTION
TASK CONTROL	CALL PTFORK (tid,tparam, subname[,arg-list])	Create a subtask and invoke subname
	CALL PTJOIN (tid-list)	Wait for completion of subtasks
	CALL PTPARAM (tparam)	Get a parameter passed to the subtask
	I = IPTSTAT (tid) I = IPTID ( ) I = IPTNAP ( )	Get status of the task Get the task identifier Get the number of available AP's
LOCK CONTROL	CALL PLLOCK (lkid)	Set the lock or wait for the lock cleared
	CALL PLUNLOCK (lkid)	Clear the lock
	CALL PLASGN (lkid[,n])	Allocate a lock
	CALL PLFREE (lkid)	Free the lock
	I = IPLSTAT (lkid) I = IPLSTATL (lkid)	Get status of the lock (and lock)
EVENT CONTROL	CALL PEPOST (evid)	Post the event
	CALL PEWAIT (evid)	Wait for the event posted
	CALL PECLEAR (evid)	Initialize the event
	CALL PEASGN (evid[,n])	Allocate an event
	CALL PEFREE (evid)	Free the event
	I = IPESTAT (evid)	Get status of the event
BARRIER CONTROL	CALL PBASGN (barid, ntasks[,n])	Allocate a barrier and set the number of tasks necessary to break it
	CALL PBSYNC (barid)	Wait until all the other tasks join
	CALL PBFREE (barid)	Free the barrier
	I = IPBSTAT (barid)	Get status of the barrier

Table 4 Macrotasking Primitives

PRIMITIVE	FORM	FUNCTION
Parallel Do	PARALLEL DO END DO	Each iteration is executed concurrently
Parallel Case	PARALLEL CASE CASE CASE END CASE	Each case is executed concurrently
Critical Section	CRITICAL SECTION END CRITICAL SECTION	Executed one at a time
Serial Section	SERIAL SECTION END SERIAL SECTION	Executed only once
Barrier	BARRIER	All tasks wait for the last one to arrive

To aid performance improvement and program debugging, various counters and functions are provided. In particular, the performance monitor is incorporated into the processor hardware and always active without any performance degradation in measuring the processor status. It includes:

Execution Counter	- Number of Instructions Executed
Vector Execution Counter	- Number of Vector Instructions Executed
Vector Element Counter	- Number of Vector Operations Executed
Vector Execution Clock Counter	- Busy Time in Vector Unit
Floating-point Execution Counter	- Number of Floating-point Arithmetic Counter

The operating system and tuning tools report the contents of the above counters to assist the analysis of dynamic profile of programs as shown in Fig. 9. This report includes execution time, vectorizing ratio, average vector length, MFLOPS and so forth. Based on these information, we can easily find which subroutine should be tuned and the causes of performance degradation. For example, the subroutine which has high percentage of the execution time and low vectorizing ratio must be concentrated on the performance tuning. And if there is serious bank conflict, then the modification of vector access pattern may be needed.

6. PERFORMANCE RESULTS

Table 5 shows various performance results of the well known benchmark tests on SX-3 system compared to the results of the former SX-2A Supercomputer. In this table, the LFK24 is a collection of 24 basic Fortran Kernels. The arithmetic mean of the LFK24 of the SX-3/14, the top end model of a single processor, achieved 553 MFLOPS while the harmonic mean is 68 MFLOPS. By the definition, the arithmetic mean usually expresses high vector performance, and the harmonic mean is governed by the slowest part of the program. Particularly, the harmonic mean sometimes represents the scalar performance only and no feature of the parallel or vector processing.

Table 5 Performance Results

PROGRAM	SX-3/14	SX-2A
LFK24 Scalar	34.4 MFLOPS	15.0 MFLOPS
Vector (A.M.)	552.7	212.4
Vector (H.M.)	68.1	25.5
LINPACK 100	314 MFLOPS	43.0 MFLOPS
TPP	4231	885.0
Matrix Multiply (1,024)	5085.6 MFLOPS	1285.1 MFLOPS

The speed of LINPACK (100×100) for the SX-3/14 is 314 MFLOPS, which is more than 5 times faster than that of the SX-2A. This increase is mainly owes to the in-line expansion function of Fortran compiler. The speed of LINPACK TPP (Toward the Peak Performance) that is allowed to modify the source code is 4231 MFLOPS, and more than 5 GFLOPS is obtained in the matrix multiply for the SX-3/14. Particularly, the speed of the matrix multiply is very close to the peak speed of the SX-3/14 of 5.5 GFLOPS.

7. CONCLUSION

In this chapter, major features of the SX-3 supercomputer system were described.

Key points to realize high performance computer system are to shorten the time for processing and to increase the parallelism. One of major factors of the former is the machine cycle time which is largely depends on device and packaging technologies. In the SX-3, 2.9 nsec cycle time is realized by high-speed VLSI chips and high-density packaging system. As for the latter, the various techniques to increase the parallelism, typically the multiple vector pipelines and multiprocessing, are employed in the SX-3. But, to utilize those hardware capabilities thoroughly, the software including the operating system, compiler, utilities and applications must be optimally incorporated. The SX-3 will exhibit high performance in a broad range of application areas by the combination of those hardware and software capabilities.

8. REFERENCES

[1] Furukatsu, T., Watanabe, T. and Kondo, R., "Supercomputer SX System with a Peak Performance of 1.3 GFLOPS and 6 nsec Cycle Time," Nikkei Electronics, 356, pp. 237-272, Nov. 1984 (Japanese).

[2] Watanabe, T., "Advanced Architecture and Technology of the NEC SX-3 Supercomputer," NATO ASI Series, F62, pp. 119-128, 1990.

[3] Watanabe, T., Matsumoto, H. and Tannenbaum, P. D., "Hardware Technology and Architecture of the NEC SX-3/SX-X Supercomputer System," Supercomputing '89, Reno, Nevada, 1989.

[4] Watanabe, T. and Iwaya, A., "The NEC SX-3 Supercomputer Series," Supercomputing Europe '90, London, 1990.

[5] Watanabe, T., "The NEC SX-3 Supercomputer System," COMPCON Spring '91, San Francisco, Feb. 1991.

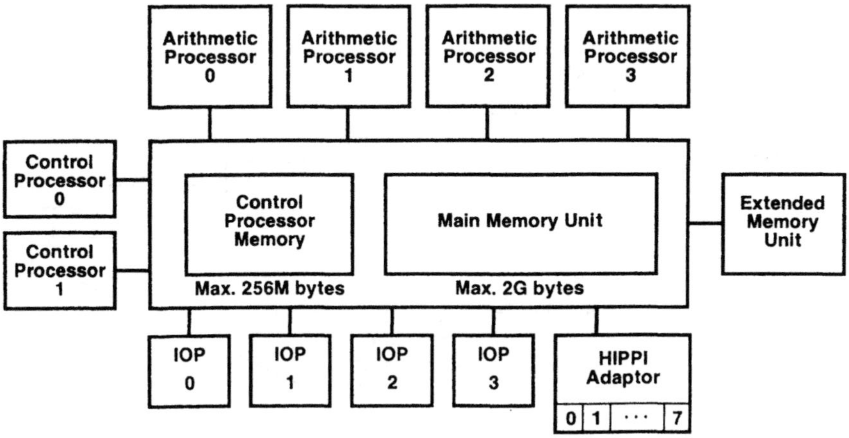

Fig. 1 System Configuration

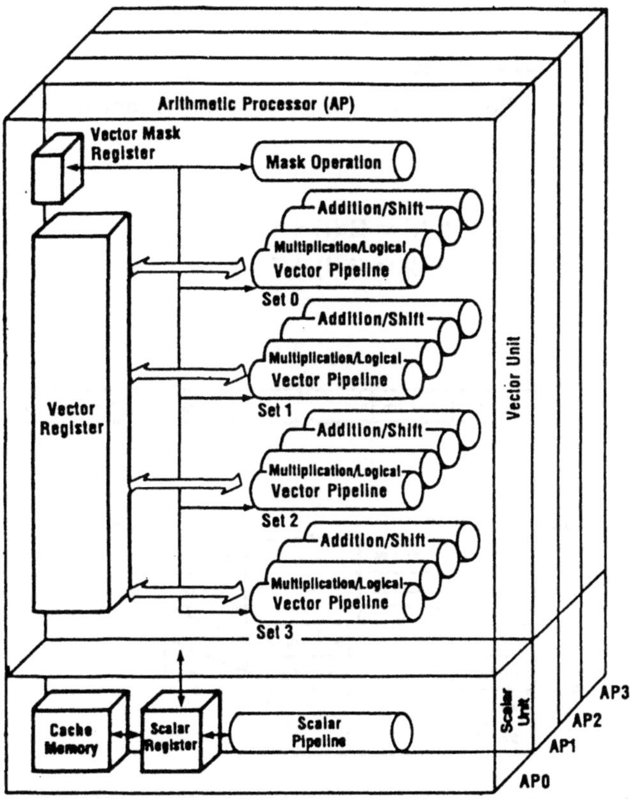

Fig. 2 Internal Structure of the AP

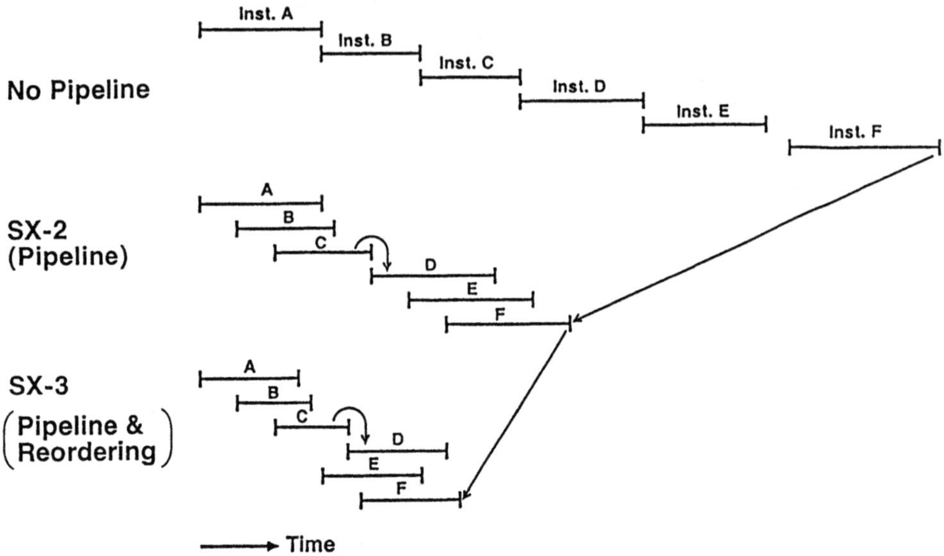

Fig. 3 Instruction Issue Control

A (I) = B (I) + C (I)
D (I) = E (I) + F (I)

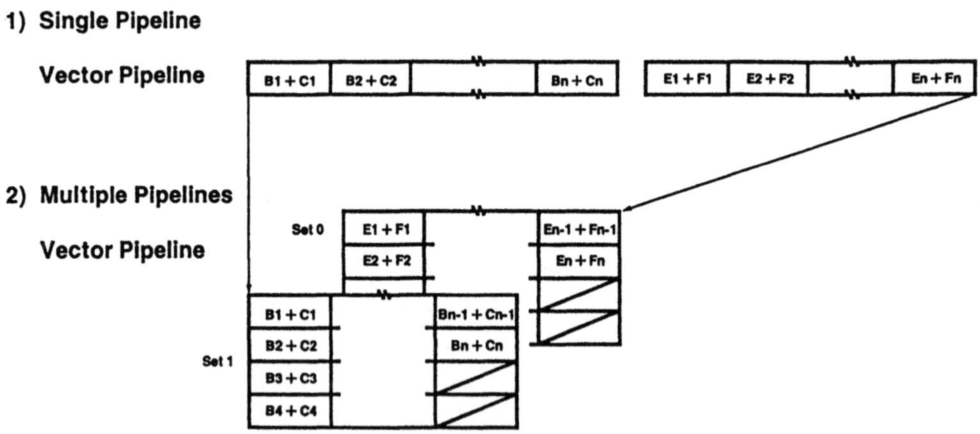

Fig. 4 Execution of Multiple Vector Pipelines

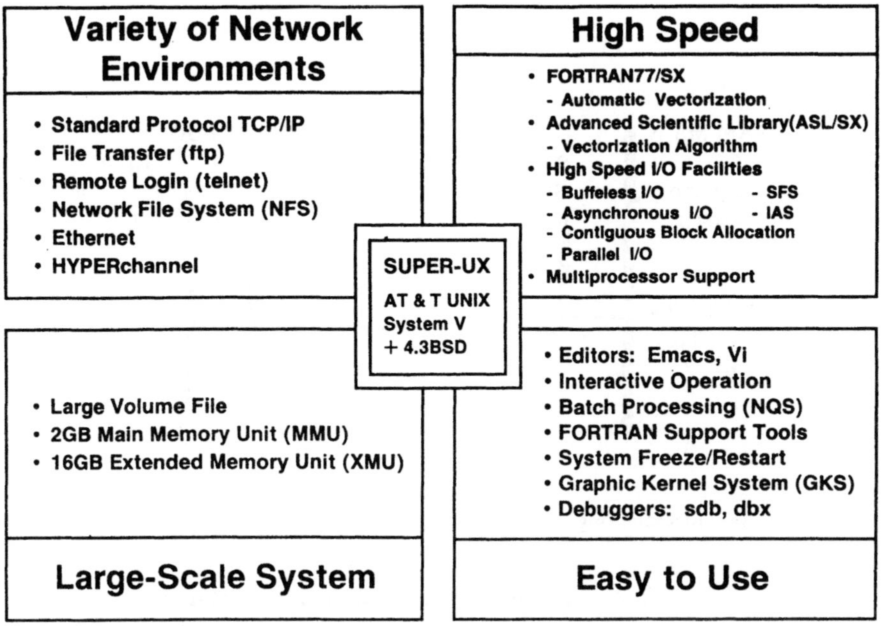

Fig. 5 Major Characteristics of Super-UX

```
                          do 195 j=1,64
5----------->             !  do 195 k=1,8
! X--------->             !  !  b(j,k)=1.00025
! !                       !  !  c(j,k)=1.00025
! !                       !  !  h(j,k)=1.00025
5-X----------         195 continue

                          do 205 j=1,4
*----------->             !  do 205 k=1,512
! W--------->             !  !  p(j,k)=1.00025
*-W----------         205 continue

                          DO 403 L= 1,MC
V----------->             !  M= (LR+LR)*(L-1)
!                         !  DO 401 J= 1,2
! 10-------->             !  !  DO 401 K= 1,LR
! ! V------->             !  !  !  M= M+1
! ! !                     !  !  !  S= REAL(K)
! ! !                     !  !  !  PLAN(M)= R*((S + FW)/S)
! 10V--------         401 !  !  !  ZONE(M)= K+K
V------------         403 CONTINUE

                          DO 10 I = 1,N
C----------->             !  DY(IY) = DY(IY) + DA*DX(IX)
!                         !  IX = IX + INCX
!                         !  IY = IY + INCY
C------------          10 CONTINUE
```

Fig. 6 Examples of Vectorized Loops

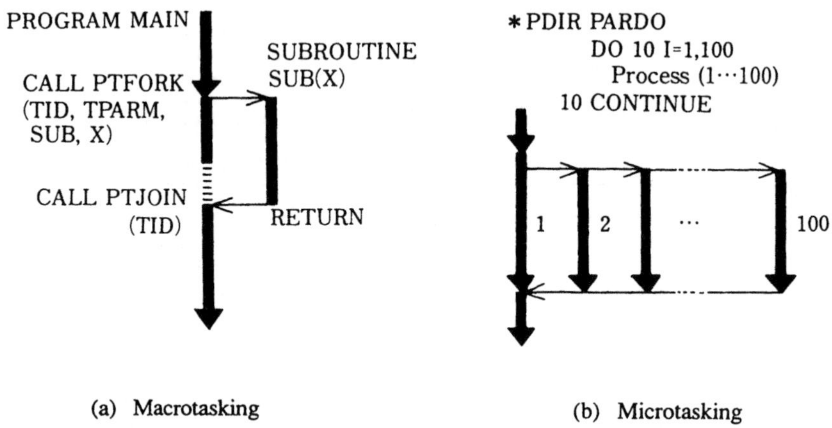

Fig. 7 Macrotasking and Microtasking

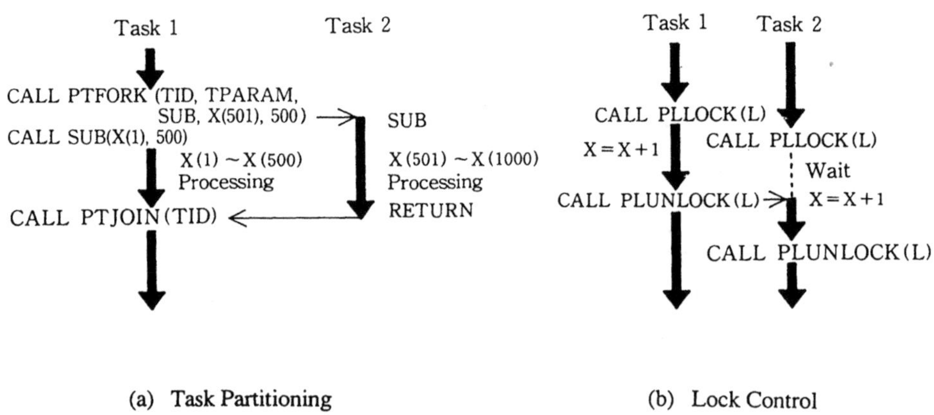

Fig. 8 Task Partitioning and Control

SUMMARY LIST

```
*-------------------------------------------------------------------------*
| ENVIRONMENT                                                             |
|       ANALYZER/SX REVISION    : REV.010                                 |
|       ANALYZED DATA           :                                         |
|                                                                         |
| PROGRAM TOTAL INFORMATION                                               |
|       EXECUTION ( CPU ) TIME =  0 : 0 ' 4 " 9  (  4009 MSEC)            |
|       INSTRUCTION COUNT      =        136839917                         |
|       VECTOR OPERATION RATIO =  98.61%                                  |
|       MOPS                   =     1445.982                             |
|       MFLOPS                 =      547.947                             |
|       BANK CONFLICT          : NONE                                     |
|                                                                         |
| OVERHEAD INFORMATION                                                    |
|       SUBROUTINE CALL FACTOR =  22.35 MICRO SEC                         |
*-------------------------------------------------------------------------*
```

ATR.	PROGRAM	FREQUENCY	TIME(%)	AV.TIME	V.OP. RATIO	AV.V.LEN.	INSTR.	V.INSTR.	V.ELEMENT	MOPS	MFLOPS	BANK CONF.
-->	ANFXRE	1	2(0.06)	2	0.75	19.0	47855	19	361	19.97	0.42	
FNC	PCGVZ	14	0(0.01)	0	0.00	0.0	7028	0	0	26.69	0.48	
SUB	PBCL	1420	76(1.92)	0	97.94	255.7	2827600	443400	113372800	1505.65	0.11	
SUB	ZKZGFG	1	14(0.37)	14	0.00	0.0	558390	0	0	37.95	14.15	
SUB	ZKZ	100	178(4.45)	1	99.32	256.0	9642500	3494400	894566400	5044.89	2349.23	
SUB	SSGGFG	1	39(0.99)	39	94.81	256.0	1153061	76800	19660800	522.27	173.65	
SUB	PSSGQ1	201	346(8.65)	1	99.42	255.0	6486624	2609591	665515492	1931.12	661.75	
SUB	PSSGQ2	201	497(12.40)	2	98.92	127.8	14230324	5939091	759127460	1542.93	527.08	
SUB	PUBGFG	1	38(0.95)	38	0.00	0.0	1328099	0	0	34.77	12.65	
SUB	PUYFXL	200	318(7.94)	1	98.07	251.0	16321400	2748400	689848400	2208.17	784.31	
SUB	OGEGFG	1	84(2.12)	84	0.00	0.0	3015323	0	0	35.48	12.39	
SUB	OGEVK	600	1522(37.96)	2	97.32	28.0	53704200	30343800	849626400	573.55	208.45	
SUB	TZGGFG	1	0(0.01)	0	0.08	5.0	12663	2	10	32.89	13.35	
SUB	TZGEL	2	299(7.46)	149	98.51	158.5	11476400	3384332	536334840	1820.23	753.67	
SUB	RZVGFG	1	43(1.08)	43	0.11	85.4	1946941	24	2050	44.84	17.98	
SUB	RZVG	10	215(5.38)	21	99.64	245.4	4006750	2122390	520937500	2422.16	1098.41	
SUB	ICRGFG	1	50(1.27)	50	0.00	0.0	1554359	0	0	30.55	13.40	
SUB	ICRAGN	400	279(6.97)	0	99.51	128.0	8520400	5220800	668363200	2402.31	881.02	

Fig. 9 Status Report of Dynamic Profile

V TRENDS IN VECTOR AND PARALLEL SUPERCOMPUTER ARCHITECTURES

Kenneth W. Neves
Boeing Computer Services
P.O. Box 24346, Seattle, WA 98034, U.S.A.

SUMMARY

The supercomputing industry is in a transitional period. The latest supercomputers are exploiting vector architecture designs which have been the basis of the industry since its inception. The inability to continue to increase single CPU performance requires increased use of parallelism. At this point in time the rapid development of powerful microprocessors has created new opportunities for very powerful, yet relatively inexpensive, massively parallel systems. In this work, a close look at the latest supercomputers is given from a software and algorithm performance perspective. Today's supercomputers are also compared with microprocessor architectures, and remarkable similarities are pointed out. Finally, important issues and considerations relative to future supercomputer developments are discussed, and software challenges identified.

1. INTRODUCTION

The supercomputer market, unlike the mid-priced air-cooled computer market, does not have a large number of players. The company that essentially established the market was CDC with their 6600 and 7600 models in the mid-60's and early-70's. The designer of these machines, Seymore Cray, established Cray Research, Inc. (CRI) which now has the largest share of the market. CDC is no longer a player in this market. US companies with products not yet on the market, include Cray Computer Corporation (Seymore Cray's spin-off from CRI), Supercomputer Systems, Inc. (partially funded by IBM, founded by Steve Chen, formerly the CRI architect of the X-MP and early Y-MP designs). Notably missing from this list are several parallel computing companies such as Intel, nCUBE and Thinking Machines. It is worthy to note that these machines, on certain benchmarks, have scored even higher ratings than the traditional vector machines. Since this work is concerned largely with vector supercomputers, we have not analyzed these machines in great detail, but they are an important part of our discussion of future architectures. These massively parallel systems have also been evolving new designs with significant changes in their characteristics. Of all these US machines, the one most dominant in the industrial market place today is the Cray Y-MP, with up to 8 CPUs rated at 333 MFLOPS per CPU for an aggregate peak performance of 2.67 GFLOPS. (1 GFLOPS is a billion floating point operations per second.) However, the three Japanese competitors offer even higher peak performance specifications.

Fujitsu, Hitachi and NEC all have had one or more generations of supercomputers in the market place over recent years. So far, Hitachi has chosen not to market its machines worldwide, at least in the US. In 1989, Fujitsu and NEC announced machines with single CPU peak performance of 4 and 5.5 GFLOPS respectively. The NEC SX-3 packages four CPUs for a combined peak performance potential of 22 GFLOPS. The Japanese companies are utilizing the same chip technology found in their more traditional mainframe products. This represents reliable technology with minimal risks in reliability, maintainability and manufacturability.

Thus, the latest generation of supercomputers boasts both US and Japanese machines competing at unprecedented computational rates (measured in terms of peak floating point operations per second). The growth of memory capacity in these systems are keeping pace with the growth of computational power.

The remainder of this chapter is organized into discussions of typical issues in designing and using supercomputers of this class, and some considerations to be prepared for in future supercomputer products. We begin with the narrow focus of looking at a single CPU of today's genre, and end with exploring the impact of RISC microprocessors and their potential use as building blocks for future supercomputers.

2. THE SUPERCOMPUTER CPU: AN OVERVIEW

A framework for discussing the similarities, differences and relative advantages of various supercomputers, requires a fundamental description of the architecture of the modern supercomputer CPU. In later sections we will discuss the use of multiple CPUs in current and future systems. In Fig. 1 a simple generic diagram of a supercomputer central processing unit is displayed. The figure itself is more a functional decomposition than a replica of a manufacturer's design. (Figure 1 will also serve in the discussion on RISC architectures in a later section.)

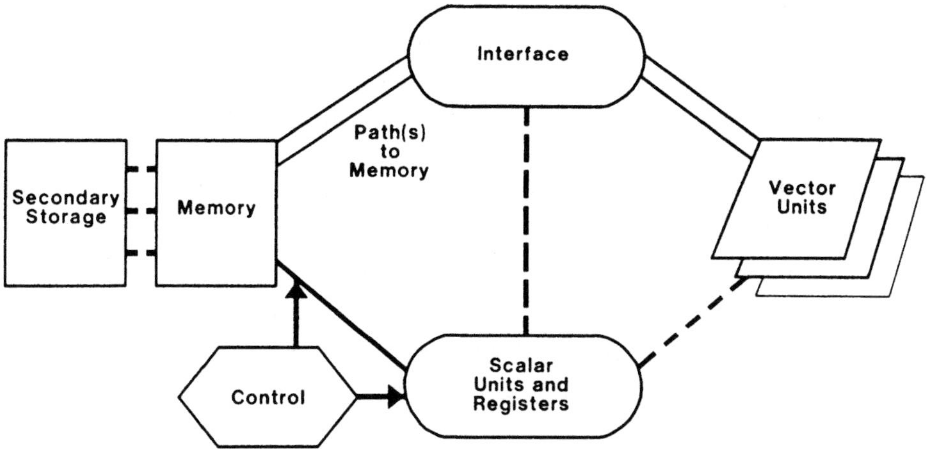

Fig. 1 A Supercomputer CPU

In, and of itself, no feature above is a sufficiently important parameter for assessing computer performance. The interplay or "balance" between these elements, however, give the computational character of a given computer. The most distinctive feature of modern supercomputers is their orientation toward processing "vectors" or arrays of elements as operands. For years the computational bottleneck in scientific computing was the processing of floating point computations. The CDC 6600, for example, tried to improve this bottleneck by using two floating point multiply units. Later the CDC 7600 exploited the pipelined concept in the functional units. Interestingly enough, the floating point units on the 7600 were, all too often, left idle. The bottleneck to computational efficiency was the rate of instruction processing associated with the overhead of fetching and storing each pair of operands and/or result. The first modern supercomputers (the CDC Star-100 and the Cray-1) circumvented this "instruction-issue" bottleneck by extending the instruction set to include vector operations. This coupled with a high bandwidth connection between the vector units and memory, through an interface (buffer or registers), characterizes modern supercomputer CPUs. In these, and subsequent supercomputer designs, one single instruction could launch a process that operated, not on one, but many operand pairs. The production rate of floating point operations became a much more meaningful measure of performance than did machine instruction rates. With the advent of the "vector" computer, the performance rating of millions of machine instructions per second (MIPS) gave way to the MFLOPS (millions of floating point operations per second) as a first order indicator of performance. We have belabored this historical digression because it will have some bearing on our discussion of RISC architectures, since these systems are or will be facing similar issues as on chip memory, cache, and floating point units compete for "chip area" which is analogous to the packaging and cooling issues found at the high-end.

The MFLOPS is now being replaced by the "Giga" (billion) FLOPS in recognition of the increased speeds of these machines. With the increased use of parallel CPUs and advances in electronics, we anticipate "Tera" (trillion) FLOPS machines in the mid to late 1990s. It is perhaps worthy to note, that even though the computational rate seems to be the paramount measure of performance, the ability to achieve this performance on any given machine is dictated by the peak sustained flow of data from memory to the vector units.

As described above, the breakthrough in performance with modern supercomputers came with the combination of pipelined arithmetic units and enhanced instruction processing allowing vector data to flow more freely between memory and the vector computational units. The critical features in this data flow process are discussed below. A more detailed and in-depth account is found in Ref. [1]. Each feature has a critical impact on algorithm design to maximize performance of a given architecture. There are no hard and fast rules, for each computer architect must balance the advantage of a given feature against the "cost" or difficulty to implement in the overall system. The key features discussed include

* the number of paths-to-memory,
* the interface between memory and the vector units,
* the allowable vector data structure in memory,
* the instruction set,
* the characteristics of the vector floating point and fetch/store units.

These features will be discussed below in the context of the vector supercomputer. Later we will revisit some of these features in the context of RISC architectures.

The number of paths-to-memory: On vector computers the operands are "vectors", large arrays of data stored in memory. In order to "operate" (e.g. add, multiply) on vectors one must fetch the vector operands from memory to the interface, and eventually store the results. A simple dyad operation requires three memory references. One must fetch two vector operands and store the result. Several computer designs, notably the Japanese machines, have multiple arithmetic pipelines. However, in these machines the multiple pipelines are logically treated as one, and the vector operands are fetched (or stored) in such a manner as to sustain full computation rates among the multiple pipeline units according to a fixed hardware scheme. The number of paths-to-memory refer to the number of simultaneous "vectors" that can be passed to and from the vector units at full computation rates. We shall observe that there are differences in the number of these "logical" paths among supercomputers, and that these differences can materially effect performance.

The interface between memory and the vector units: The interface between memory and vector units is usually a vector register set. Notable exceptions were the CDC/ETA Systems computers which used a buffer, and the IBM 3090/VF which uses an associative cache. Registers are supported by the vector instruction set, and vector data is passed modulo the length of the vector registers. For example, on the CRI machines this length is 64 words. Compilers or application software cannot utilize buffers as temporary resting places for accumulation of data or storage of interim results. Thus buffer oriented systems suffer from requiring increased memory traffic over register or cache based systems. Other types of interfaces include caches, local memories (essentially a cache under user control, e.g., Cray-2 local memory), and scalar registers. With the recent demise of ETA Systems, all the high-end supercomputers are vector register oriented. The use of cache based systems may very well reappear in supercomputer designs. The fundamental difference between registers and a cache rely in the responsibility for data transfer. Cache based systems usually have an "automatic" system for bringing data into the cache. This is often based on a set of pre-defined rules which attempt to minimize data flow based on a number of factors. Register oriented machines are directly programmed to fetch/store data between memory and registers as dictated by the user (compiler). While supercomputers are almost all register based, Convex and Alliant, two mid-priced mini-supercomputer companies, have used caches to some success in their designs.

The allowable vector data structure in memory: To a mathematician or scientist, a vector is simply an ordered array of numbers. For a computer, this array doesn't exist until it is designated by the memory location of its ordered components. Various supercomputers have limitations as to the flexibility allowed for the specification of vectors. Machines that do not employ vector instructions, can avoid this complexity. There are three common complexities found in manipulating vectors stored in supercomputer memories:

1. Contiguously stored vectors (stride 1),
2. Regularly stored vectors (allowing an arbitrary, but fixed, stride between array components as stored in memory),
3. Randomly stored (allowing successive vector components to be designated according to a list or index vector – a form of indirect addressing).

The random storage of vectors, when supported by a sufficiently rich vector instruction set, provides very useful capabilities across a broad range of linear algebra and other mathematical algorithm classes. While all supercomputers generally support type 1 above, not all support type 2 and 3. The CDC/ETA machines, for example, support only type 1 in all of their vector operations. Today most other machines support all three types, yet some of them degrade in full performance on type 3 operations. The Cray-2 for example, performs at one quarter of the peak vector computation rate on random stride vectors. [Note: items 2 and 3 above can cause cache systems to suffer if the random or stride arrays are out of the bounds of the contiguously stored data in cache.]

The vector instruction set: As pointed out above, vector instruction processing is the distinguishing feature of supercomputers from scalar computers. The performance potential of modern supercomputers can only be achieved in "vector mode", i.e., utilizing vector instructions. A full complement of vector instructions includes dyad and triad operations (the latter being linked add/multiply operations). These instructions are to be supported on all allowable vector types, contiguously, regularly, and randomly stored. While all the latest machines support these constructs on basic operations, some degrade with non-contiguous storage. For example, Fujitsu's VP-200 and VP-2600 can degrade from peak performance when fetching pairs of non-contiguously stored vectors. Some machines have more vector instruction types than others, but usually the extra instructions do not impact mainstream scientific computations. (There are significant exceptions, but the details are beyond the scope of this article.)

The characteristics of the vector floating point and fetch/store units: This area has become most critical in rating performance of supercomputers. With the advent of parallel CPUs, deficiencies in this area can be even more pronounced. Pipelined units work like "mini" assembly lines. The number of pipe segments are analogous to "work stations" in the assembly line. The more segments, the longer it takes the assembly line to fill up and the longer the start up time, or time to produce the first result from a dead start. This "wait" time is the overhead of vector computing. Memory fetch/store units and arithmetic units all have pipe segment latency which must be considered in performance evaluations.

3. A SUMMARY OF SUPERCOMPUTER HARDWARE CHARACTERISTICS

The above features, along with computer memory, comprise the basic tools that compilers and application programs use to achieve high performance. Information on the newest Japanese machines can be found in this volume. Yet, a fairly complete comparative picture is often elusive due the fact that expositions on a single machines are often done without regard for comparative analysis. The statistics given in this section are largely developed from material provided by vendors at the time of market introduction. Two things often change quickly in a machine design: cycle time and memory size. *The reader therefore may wish to compare cycle time figures and memory size information against current manufacturer literature.* The subsequent results can then be scaled by cycle time if it has changed, and memory growth has obvious benefits. Another matter concerns memory speed. As memory technology improves (usually through bigger memory chips) the access time to memory will improve. This usually has the immediate effect of improving the start-up time (or time to achieve first word fetch). A second effect will be the "staging" or "segmenting" of the memory load/store pipeline. This will have an immediate impact on performance of shorter vector performance as will be described.

Taken alone, peak performance ratings are very misleading indicators of performance. Peak performance is an indication of what computational rate a given processor is guaranteed not to exceed. Taken in this light these rates (often dubbed "macho FLOPS" by supercomputer hackers) provide an indication of underlying asymptotic performance potential for very ideal operations of vectors of enormous lengths. Table 1 below gives the single CPU performance of various machines, and where applicable includes multiple CPUs ratings. All the data is for 64-bits precision.

Table 1 Peak Performance Rates

Computer	Cycle Time in Nanosec.	Single CPU (Multiple) Peak MFLOPS Rating	
CRAY-1	12.5	160	
CRAY X-MP	8.5	233	(932, 4-CPUs)
CRAY-2	4.1	488	(1952, 4-CPUs)
CRAY Y-MP	6.0	333	(2666, 8-CPUs)
CRAY-3	2	1000	(16000,16-CPUs)
CYBER 205	20	200	
ETA-10/G (1988)	7	625	(5000, 8-CPUs)
FUJ. VP-400E	7	1700	
FUJ. VP-2600	4	4000	
HIT. S-810/20	14	630	
HIT. S-820/80	4	2000/3000	
NEC SX-2	6	1300	
NEC SX-3	2.9	5500	(22000, 4-CPUs)

Main memory is an important characteristic to memory constrained applications. Many important computations in science and industry are constrained by memory and storage bandwidth. Computational fluid dynamics (CFD), large structures problems, electromagnetics, seismic analysis, medical imaging, are all examples of applications whose model sizes are constrained by memory size.

Memory sizes from all vendors are growing. We expect to see main memory sizes of several Gigawords by the mid-90s. In massively parallel machines aggregate memory will be one to two orders of magnitude greater.

Paths-to-Memory: A quick listing of this characteristic among vendors (see table below) reveals that this feature continues to be critical. Anything less than three paths-to-memory, puts a greater burden on algorithm design or compiler optimization efficacy. Many algorithms in use today have inner-loops that require at least two fetches and one store. Often these algorithms can be recast to provide less demanding inner-loop memory traffic, but all to often this is not done, and in some cases cannot be done.

Table 2 Main Memory Sizes (64-bits words)
(at introduction)

Computer	Size MW	No. of Banks	Secondary Mem. Size
CRAY X-MP/4	8	64	512
CRAY-2	256	128	none
CRAY Y-MP/8	32	256	512
FUJ. VP-400E	32	256	none
FUJ. VP-2600	256	256	1024
HIT. S-820	64	128	1500
NEC SX-2	32	512	256
NEC SX-3/44	256	1024	2048

One path-to-memory has often proven to be too restrictive. The Cray-2 has only one path as did its predecessor, the Cray-1. It was the upgrade of this feature in the Cray X-MP that made a big difference in its performance. In fact, as reported in Ref. [1], an optimally coded triad operation on the X-MP is four times faster than an optimally coded triad on the Cray-1, despite the X-MP boasting only a 25% faster cycle time. This improved performance is solely due to the extra paths-to-memory. Later CRI designs, the Cray-3 and Y-MP both have three paths-to-memory. Many basic computational loops in FORTRAN algorithms are severely handicapped by only one path-to-memory. As indicated in the below table the Fujitsu VP-200 had 2 paths-to-memory on contiguously stored vector operands, but for stride or random vectors, it has only one path. The VP-2600 is similar to the VP-200 in paths-to-memory. The VP-400 added more pipes (4 sets) which caused double the number of words per cycle to be fetched to support peak speeds. This reduced the machine to a single logical path for all vector operations. So, the number of paths-to-memory has not been consistent among vendors, within a single vendor's line, or even (in the case of the Fujitsu VP-200 and 2600) within a single machine.

Table 3 Paths-to-Memory

Computer	#of Logical-Paths-to-Memory per CPU	Latency in Cycles
CRAY-1	1	11
CRAY X-MP	3	14
CRAY-2	1	35-50
CRAY Y-MP	3	17
CYBER 205	3	50
ETA-10	3	not available
FUJ. VP-200	1 or 2	31-33
FUJ. VP-400	1	31-33
FUJ. VP-2600	1 or 2	
HIT. S-820	2	not available
IBM 3090/VF	– associative cache –	
NEC SX-2	4	not available
NEC SX-3	3	60-70

4. PARALLEL VECTOR COMPUTATION, AND LATENCY IN DESIGN

Perhaps the most striking realization in reviewing supercomputer architectures (at the high-end), is that all manufacturers confront the same barriers of time and space. The only alternative is parallelism. In fact, all manufacturers are employing about the same amount of parallelism in their designs, but in different forms. For example, the Cray X-MP 4-CPUs machine has two floating point vector units per CPU (an add and a multiply). Thus, the whole system has 8 floating point units. It is termed a parallel CPU machine because it has 4-CPUs. On the other hand the NEC SX-2, introduced as a competitor to the X-MP, is a single CPU machine. Yet, within the CPU it has 4 add pipes, and 4 multiply pipes. In terms of floating point operations, the two machines have the same degree of parallelism. The architects must manage a steady flow of operands to about as many pipes during peak performance. To appreciate the differences between the two machines one must look first at the bottlenecks, latencies, and flow of vector computation, and the likely use of parallel CPUs.

The latest generation of machines from the Japanese manufacturers have followed their earlier trend of using multiple pipelines within a single processor to achieve greater peak performance. In order to do this, they must provide greater bandwidth between the CPU and main memory. For example, in order to support, four multiply pipes producing one result per cycle each, a stream of 8 operands must be available each cycle after start-up to support two logical paths (i.e. a fetch of two vectors). The Hitachi S-810, the NEC SX-2 and the Fujitsu VP-200, 400, all have multiple pipelines as indicated below. The CRI series of machines have consistently provided only one add and one multiply pipe. In order to achieve higher performance rates, more CPUs were added.

Table 4 Floating Point Pipes Per CPU (Earlier Machines)

	Number of Pipes/CPU
CRAY Y-MP	2
FUJITSU VP-400	8
HITACHI S-810	8
NEC SX-2	8

The second generation of Japanese supercomputers employ many more pipelines per CPU. It is not really known how many floating point pipes make a well balanced vector/parallel architecture. Many believe one add and one multiply per CPU is not enough, and others argue that 16 is excessive. The answer, of course, is application dependent. Table 5 indicates what the Japanese have done in their more recently announced machines.

Table 5 Floating Point Pipes per CPU (New Machines)

	NUMBER OF PIPES/CPU
FUJITSU VP-2600*	8
HITACHI S-820*	8
NEC SX-3	16

* These designs use a combination add/multiply pipe that is chained or linked, allowing an add result to be input to a multiply. This achieves two floating point results per cycle per pipeline, for chained operations. This is in contrast to NEC which provides fully independent add and multiply pipelines.

The Fujitsu VP-2000 series has taken the following approach to improving the multiple pipeline computational output. In earlier Fujitsu designs, as already observed, providing the proper bandwidth to memory (as evidenced by reduced paths-to-memory) has been a design problem (i.e. a bottleneck in data flow). Adding pipes exacerbates this design problem for the architect. In the VP-400E, an approach was taken to create a linked pipe. This was an add/multiply pipe that could perform a vector multiply followed by a vector add with out going back to registers with the intermediate result. This could be considered hardware "chaining" or "linking" of the multiply and addition operations. If only a multiply is desired, the add function is not performed, but the pipe length remains as long as the chained operation. On the VP-2600, this concept was extended as follows.

The VP-2000 series has the following floating point units - 2 add/multiply units, and one divide unit. This could be called a pipe set. On the VP-2600, there are four such sets. Each arithmetic unit is composed of four pipelines which are run simultaneously. The add/multiply pipes are hardware "chained" to produce linked add multiply results. Therefore, the VP-2600, in theory, can perform 8 vector adds chained with 8 vector multiplies per cycle. This results in 16 floating point operations every 4 nanoseconds. The divide cannot execute while all the add/multiply pipes are going. Thus, one arrives at the 4 GFLOPS peak performance figure. An optional feature of the VP-2600, is the addition of a second scalar CPU, the model VP-2600/20. The rationale for this feature is that most applications cannot keep a vector CPU busy. A second scalar CPU will improve job stream throughput and allow sharing of the vector CPU. This may find usefulness in mixed business scientific environments.

NEC seems to have pushed the multiple pipeline concept to the extreme in the SX-3. This machine has 8 multiply and 8 addition pipelines, for a total of 16 floating point units per CPU. Unlike Fujitsu's VP-2600, these are independent units, not requiring chaining to attain peak performance. This is double that of the SX-2. With this many pipelines to fill, a much higher bandwidth has been established between registers and main memory. To support this feature the pipelines are grouped into two logical sets of 4 add/multiply pairs each. Each logical set is operated much like the SX-2. A vector operation is automatically spread among the 4 pairs of pipelines as appropriate. To achieve both logical sets functioning simultaneously, the instruction process supports issuing overlapped instructions. That is, a vector instruction takes two clock periods to execute, but a second vector instruction can be initiated one cycle after the first. Subsequently, two vector operations will proceed simultaneously, each using a separate logical set of pipes. The net result is 16 floating point operations per cycle. Both instructions are fully supported by two vector fetch and one vector store pipelines. This means that 3 logical paths-to-memory are fully supported. With a 2.9 nanoseconds clock, the NEC SX-3 (called the SX-X in the US) is a 5.5 GFLOPS/CPU design – in its 4-CPUs configuration, it is a 22 GFLOPS peak performance machine. There are some further subtleties, such as not all paths on all CPUs can be active simultaneously.

At first glance one might say that all of these decisions about numbers of pipes, CPUs, and chaining (linking) are somewhat arbitrary. They are all an attempt to create greater performance within the constraints each designer has. Unfortunately, peak performance is not what users want to optimize, but sustained performance on an application. If you utilize many pipes, you increase the length of vectors required to achieve good performance. If you utilize more CPUs, the users must figure out parallel algorithms that minimize synchronization and other overheads. Offering linked add/multiply pipes produces a need to organize computation (either by hand or by compiler) to achieve a greater percentage for linked operations in applications. For every decision there is a down-side for application developers and users. This is the price one pays for the unabated increase in potential performance the industry has enjoyed for several decades. It is worthwhile to consider some of these matters in more detail, which we do in the next section.

5. A STUDY OF VECTOR START-UP TIME

The use of the pipelined technique in vector load/store operations can be a great equalizer. For example, a manufacturer with slower memory technology (longer delay to retrieve data) can pipeline the memory fetch operations with more segments (stages) and achieve a high effective bandwidth for large amounts of data. That is for long vector operations, the resulting rates can be as good as another manufacturer who may have faster memory. The problem with this approach occurs when vectors are not "large". What is large? R. Hockney and C. Jessophe[2] defined a term useful in addressing these often fuzzy latency issues. The term is called $N_{1/2}$ (read N sub a half). The term will be redefined here, and used in extending a latency result in Ref. [1] for single pipelines to multiple pipeline operations. This will enable us to better compare multiple CPUs designs of CRI with the multiple pipelined CPUs of the Japanese vendors.

> *Given an operation (or even an algorithm), $N_{1/2}$ is defined to be the length of a vector to achieve one half the asymptotic peak performance for the given operation (or algorithm).*

Perhaps the best way to describe $N_{1/2}$ is graphically. Figure 2 displays the performance in MFLOPS for the vector triad, A*X+Y (a Scalar, "A", times a vector "X" Plus a vector "Y", called SAXPY), as a function of vector length. The curve is obtained by modeling the time of a vector operation as follows:

$$T = S + K * N,$$

where S is the start-up time to fill the pipeline, and K is a constant (related to the pipe cycle time). Solving this equation for computation per unit of time yields,

$$N/T = 1/(S/N + K).$$

As N approaches infinity, N/T approaches the asymptotic rate, R, for the vector process, as indicated in Fig. 2.

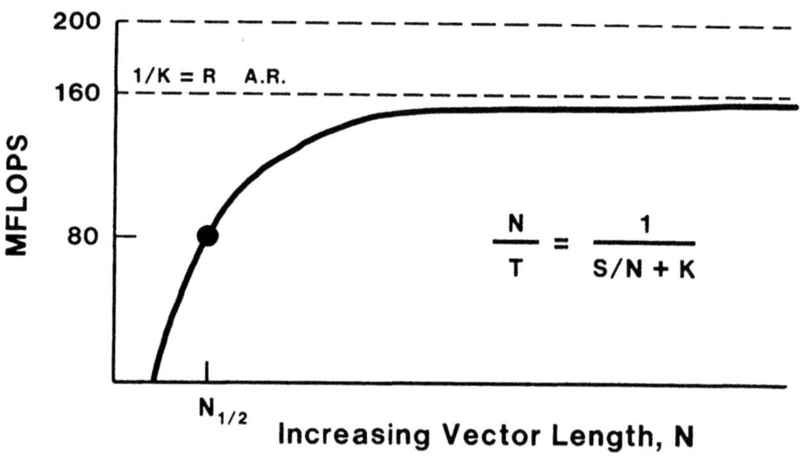

Fig. 2 $N_{1/2}$ **for a Vector Operation**

The value of the $N_{1/2}$ parameter is largely heuristic. While peak performance gives one an ideal of potential performance on long vectors, $N_{1/2}$ reveals performance in less than ideal situations. If performance were being measured on vector operations of length equal to $N_{1/2}$, then performance would be, by definition, 50% of peak. $N_{1/2}$ gives us a "feel" for what is a short vector and what is a long vector. A common heuristic to apply is:

> *If vectors are very much shorter than $N_{1/2}$, performance is very poor. While vectors of length twice or three times that of $N_{1/2}$, usually operate significantly close to the peak performance.*

One consequence of this heuristic parameter is that what is a "long" vector on one machine, might be a "short" vector on another, for $N_{1/2}$ is a function of peak speeds and start-up time for vector operations. It is therefore worthy to study $N_{1/2}$ in more depth. In reference [1] it was pointed out that multiple CPUs generally double the asymptotic peak rates as well as the $N_{1/2}$ values. The following result was also established in that reference:

> *If an operation is the result of an M-segmented pipelined process with one result per cycle, then $N_{1/2}$ is simply equal to the number of pipe segments M.*

The number of segments in the pipe has been called the "depth of parallelism" (defined in Ref. [3]) of a vector process, for at any instant in time a full pipeline is computing (at some stage) M operations simultaneously. This result, and the definition of $N_{1/2}$, make it a computable parameter from hardware specifications. In addition, given an operation or algorithm, one can experimentally estimate $N_{1/2}$ by plotting FLOPS rates versus various veetor lengths and read an approximation to $N_{1/2}$ off the graph. For machine instructions such as simple vector adds or multiplies, one can use the above result, and add the number memory fetch/store segments to the number of floating point unit segments. The resulting sum is $N_{1/2}$. In multiple pipelined machines, or in multiple CPUs machines, which produce more than one result per cycle, the above result is extended as follows.

Assume that a vector operation achieves a rate of P floating point results per machine cycle. This could be due to having P parallel pipelines sharing the chores by chaining or parallel processing the vector stream. Also assume that the vector operation is composed of a memory reference pipe chained with the floating point unit pipe so that the resulting combined segmentation is M segments long. Let C_T be the cycle time, then the time to perform a vector operation of N floating point results is given by

$$T = C_T(M + N/P).$$

For a vector of length N, effective FLOPS rate is given by,

$$N/T = N/[C_T(M + N/P)],$$

or

$$N/T = \frac{1}{C_T(M/N + 1/P)}.$$

Allowing N to approach infinity, results in the expected asymptotic rate of P/C_T. Half of this performance rate is $P/(2 * C_T)$. This is achieved when $N = P * M$. Thus, $N_{1/2} = P * M$. Simply stated:

> *For multiple pipelined operations (either chained pipes or duplicated pipes) that achieve P results per machine cycle, $N_{1/2}$ is equal to the total number of segments in the combined pipelined operations considered, times P.*

We now have a full set of tools to analyze vector latency in supercomputers. If the hardware is available, experimentation can be used for complex operations or algorithms to establish $N_{1/2}$. For those any machine whose specifications are known, one can estimate the $N_{1/2}$ value for simple vector operations from the number of segments in memory pipelines, segments in the floating point pipelines, and the number of results per cycle.

Table 6 below gives $N_{1/2}$ values estimated experimentally for several machines from Ref. [4]. The FORTRAN SAXPY operation is used which is scalar times a vector with the result added to another vector. This operation can be chained on machines that allow chaining.

One can compute the above result from hardware design information. For example, using the Cray X-MP, where there is a 14 cycle wait time for read, 13 segment chained add/multiply, with 2 results per cycle. We can expect an $N_{1/2}$ of 54. One should also expect a peak rate of close to the maximum of 210-230 depending on the model. The FORTRAN results, in the table show a smaller $N_{1/2}$, and a smaller peak rate than theoretical. However, careful assembly coding can achieve the higher peak. If start-up time remains constant, the higher the peak rate, the larger the value of $N_{1/2}$. Reference [1] reports an experimentally computed $N_{1/2}$ value of close to 80, and peak of over 200, from Cray assembly coded loops. The theoretical model above does not consider the fact that all vector operations are modulo the length of vector registers, introducing some additional overhead and increases to theoretical $N_{1/2}$ values.

Table 6 Values for $N_{1/2}$ (FORTRAN SAXPY)

Computer	$N_{1/2}$	FORTRAN Peak Performance
CRAY-1	20	45
CRAY X-MP	37	101
CRAY-2	30	55
CYBER 205	238	170
FUJ. VP-100	200	140
FUJ. VP-200	120	190
IBM 3090/VF	34	53
NEC SX-1	30	240
NEC SX-2	80	575

Another example illustrates how computing theoretical $N_{1/2}$ values can depend on often obscure information. The Fujitsu VP-200 has a 31 segment read pipe (33 for non-contiguous data). The 2 add and 2 multiply pipes have 6 and 7 segments each. Assume the scalar value for the multiply is in a register, and assume that the vector to be added is in the register as well. (This is not unreasonable since in Gaussian elimination, where this operation is often used, one can accumulate the inner-loop results in a register via column-ordered elimination [1].) After 31 cycles the first pair of fetched elements reach the registers. After 14 more cycles the first quartet of results are computed. The result rate is 4 per cycle (2 adds and 2 multiplies) after a total 45 cycle start-up. According to the theoretical result

$$N_{1/2} = 45 * 4 = 180.$$

The experimental result was 120 and the theoretical result is 180. Disparities of this kind in $N_{1/2}$ values computed experimentally occur between assembler coded loops and FORTRAN. FORTRAN typically achieves lower peak rates, and consequently, smaller $N_{1/2}$ values.

Another factor, not considered in the theoretical calculation, is the use of overlapping operations. It is possible to issue the next vector instruction in a sequence, before the first is completely done. This greatly reduces the start-up time of a vector operation since the pipeline filling process can be shared with the "emptying" of the previous vector operation. The Fujitsu VP-2600 can overlap as many as four vector instructions. Thus, in a series of vector operations, the segmented read pipe need not completely empty itself before a successive read can take place. This judicious scheduling could have the effect of reducing the wait time of the successive vector read operation. NEC machines do this as well.

For this reason, another parameter can be defined, the

> *Sequenced $N_{1/2}$: The vector length to achieve half of peak performance for a vector operation within a sequence of vector operations.*

Unfortunately, the theoretical calculation of $N_{1/2}$ (sequenced or regular) is mathematically simple, but in practice obtaining the proper hardware specifications is quite difficult. As the machines become more complicated, the exceptions of overlapping rules, pre-fetching schemes, and a host of other details exacerbate the problem. For this reason when the hardware is available, the experimental determination of this useful parameter becomes the most fruitful approach.

The remainder of this discussion will be focused on the following machines: the Cray Y-MP, NEC SX-3, Hitachi S-820/80, the Fujitsu VP-2600. While we have observed that the theoretical estimates of $N_{1/2}$ values can differ from FORTRAN, they do serve as a comparitor of latency in vector operations when benchmarks are not available. The first ingredient in this computation is the memory latency measured in cycles of delay until first word availability. The second is the number of results produced per cycle once all the pipes are full. Finally, one must pick an operation or computation to analyze. We shall select the contiguously stored SAXPY operation discussed above. However, we shall make the following assumptions:

* The operation is performed once.
* One vector operand is stored in the registers.
* The final result will not be stored (simulating an accumulation for a next iteration).

Table 7 lists the latency of the fetch, add and multiply pipes measured in cycles (pipe segments). This is information to be used in the calculation of the theoretical $N_{1/2}$ listed in Table 8. Unfortunately, the values F, H and N, are considered proprietary at the time of this writing. In the past the Japanese vendors have had comparatively large memory latencies when measured in cycles of delay for initial components of a vector load to register. However, one can estimate the values of F, H and N.

Table 7 Vector Operation Latency

COMPUTER	Mem. Refs. Latency in Cycles	Floating Pt. Latency in (Add + Mult.)
CRAY Y-MP	20	6 + 7
FUJITSU VP-2600	F	11
HITACHI S-820	H	6 + 8
NEC SX-3	N	8 + 8

After the pipeline is full, a number (equal to the length of the vector register) of operations are completed (one result per cycle per pipe) until the vector register length is exhausted. The next set of elements is then processed until the entire vector length is exhausted. The subsequent set is usually fetched in overlapped fashion so that there is no repeated pipe start-up time (i.e. a pre-fetch has occurred). There may be a several cycle gap between sets. For the Cray designs the register length is 64 words. For the Japanese machines these vector "sets" are generally larger, but so is the overlapped latency. This effect will be ignored for all machines. The table below lists the number of results per cycle achieved, and the estimated theoretical $N_{1/2}$ (for a single CPU).

Table 8 Theoretical $N_{1/2}$ (Linked triad, with one vector fetch)

Computer	Results/Cycle	$N_{1/2}$
CRAY Y-MP	2	66
FUJITSU VP-2600	16	$16*F$
HITACHI S-820	8	$8*H$
NEC SX-3	16	$16*N$

In the past, the primary reason the Japanese entries have such large $N_{1/2}$ values, is the large memory latency. While this can be shortened considerably in a sequence of overlapped instructions in the newer designs, it still is of concern. Based on published memory speeds and past machine designs we estimate that F, H and N will range from 70 to 100 on an initial vector operation.

For the sake of discussion, assume that F, H and N are on the order of 70. With this value of latency, the $N_{1/2}$ values for Fujitsu's VP-2600 and NEC SX-3 would be on the order of 1120. Using these theoretical estimates, and the heuristic rule of thumb, one can observe that, to be efficient in vector operations, one must have vector lengths on the order of 2000 to 3000 for the these machines, while only on the order of 150 to 200 for the Cray machines. One must understand that greater peak performance generally will bring these larger latencies. However, unless problem sizes and, consequently, vector lengths grow with CPU power, less efficient use of the hardware will result.

REMARK: *One should note the very exacting assumptions used in Tables 7 and 8. For memory-to-memory operations (i.e. requiring the final vector store) machines with path deficiencies would suffer lower peak performance, and require even longer memory delays than listed in these tables. For example, a two path machine would require a N-cycle wait plus start-up time for the final store operation, effectively reducing peak performance by a factor of two. On the other hand, three path machines would be able to almost completely overlap the start-up time of the store operation to achieve full performance. Similarly, if two vectors had to be fetched (instead of assuming that one was already in the register), the one path-to-memory architecture would perform much worse.*

6. PARALLEL COMPUTATION

All of the manufacturers represented in Table 8 have indicated intentions to offer multiple CPUs machines. Reference [5] details how compiler technology today, and for the near future, can exploit parallelism. There are a number of broad issues in parallelizing application programs. At the loop level, a form of parallelization is easily implemented by the compiler. There are two possibilities. First, a nested loop can be divided among CPUs at the outer loop level giving each CPU a sequence of inner-loop vector operations. This is the most advantageous way to deploy the multiple CPUs at the loop level. Vector lengths remain the same, and the number of vector operations remain the same. The overhead of parallelization is the only penalty. Unfortunately, not all loops are nested, and not all that are nested can be recognized as being independent and allow for this technique.

A second possibility is that a single loop is segmented and spread among available CPUs. This technique results in the length of each loop being divided by M, the number of processors used. This has two effects. First, what was once a single vector operation, becomes M vector operations. Second, each vector operation is using a shorter vector. In this situation what was considered a long vector for a single CPU, now must be M times longer. That is, what was a adequately long vector length for efficiency on single CPU vector operations may not be long enough to be efficient on multiple CPUs machines. Consider an 8-CPUs computer. If a single CPU vector operation had an $N_{1/2} = 150$, then 450 might be considered adequate length to achieve efficiency on a single CPU, while a vector length of 3600 is necessary for a single un-nested loop implemented on 8-CPUs. Moreover, this process doesn't scale. That is, an 8-CPUs machine cannot necessarily handle problems uniformly scaled to be 8 times bigger unless its memory and other resources are enlarged. Further, in order to achieve longer vectors (often one dimension in a multi-dimensional problem) memory might have to grow with the square or cube of vector length.

It is quite possible, that as parallel CPUs are exploited, the efficiencies enjoyed by vector CPUs will decrease for lack of algorithmic approaches that avoid the degradation of the second scenario above. In other words, the current trend to automatic parallelization must be used with an awareness that parallelizing vector instructions can produce less efficient vector operation in the individual CPUs. Parallelization from the top duplicates entire processes and offers the opportunity to gain the advantages of parallelism without the needless sacrifice of vector efficiency.

While $N_{1/2}$ values are generally increasing, so are the size of memories. The NEC SX-3, for example, requires longer vectors to be efficient than other designs, yet it provides large enough memory and secondary storage to support more complex computations with, perhaps, longer average vector lengths. Multiple pipeline machines with very high computation rates will probably be competitive on smaller problems. Both Fujitsu and NEC have attempted to ameliorate their short comings in start-up times with techniques for overlapped instructions, appropriately placed caches, and large register sets. Finally, analyses such as these focus so much in efficiency (in terms of achieving high percentage of peak performance), one can forget that having very high performance peak numbers can allow fairly fast computations *even at low efficiencies.*

7. RISC ARCHITECTURES

RISC-based microprocessors have captured the interest of several industries, and in fact, lead an industrial revolution of their own. The technology is driven by the volume market of consumer electronics. It was once thought that memory chip design fostered the innovation of the semiconductor industry, and that microprocessors inherited those advances. Intel, Sun, MIPS, Motorola and others have shown that this is simply not the case. Innovative manufacturing, design and device densities are shared in both these semiconductor industry sectors. With workstations, consumer electronics, industrial use in robotics, autos, planes and an enormous number of other areas, microprocessors are a high volume product. Given this financial base, research and innovation is nurtured.

It would be far beyond the scope of this discussion to delve into these chips with as much detail as we have with supercomputers; yet so much of the previous discussion is relevant to their architecture, we would be remiss not to comment. Further, there are at least two and, indeed we predict, many more massively parallel computer companies who are basing their designs on the commodity RISC market for building blocks in their system. This is for good reason.

The performance curves of virtually all RISC chips over time is increasing with greater slope than any other sector in the industry. While there are technological reasons why this cannot keep up forever, one can see another 3-5 years improvement due to many factors. In many ways the modern RISC architectures look very much like a supercomputer CPU on a chip. In their current stage of evolution most have not had to rely on vector instructions to keep their arithmetic units fully busy serving data from on-chip cache. Most of these microprocessors employ only scalar instructions and some have single cycle floating point units. An example of a popular scalar RISC architecture is Sun's SPARC. The SPARC 2 with its 40 MHz clock is rated at 25 MIPS (millions of instructions per second) and 4.2 MFLOPS. This is a very typical ratio of floating point to MIPS performance found in scalar processors. Yet, if one examines a microprocessor that really pushes floating point performance such as the Intel i860, one can see the exploitation of every architectural "trick" found in high-end supercomputers and air-cooled mid-ranged mini-supercomputers. To illustrate we examine the i860 from the perspective of Fig. 1.

i860 Memory: This is 4-ways interleaved, 64-bits memory.

Interface between Memory and Floating Point Units: There is a bus interface (data path) to the FPU control unit, but there is also both a 8 Kbytes data cache as well as vector registers. This interface exploits almost all known mechanisms!

Floating Point Units: Separate add and multiply pipelines, each 3 segments long. Aggregate peak rate of 66 MFLOPS.

Instruction Processing: Separate code cache for instructions from memory. The instruction set includes "pipelined" instructions which are the equivalent of vector instructions. In fact, in obtaining data from cache to feed the FPU registers can be done in "Cray mode" as it is heuristically called.

Other Features not in Fig. 1: Graphics processor which can execute both scalar and vector graphics instructions.

In looking at these features, one can say that the i860 looks very much like a supercomputer on a chip. To anyone who has programmed in assembly language on vector computers and array processors, the i860 looks very similar, only more complex. It is not surprising to hear that initial compilers for this machine were not as efficient as they could be, for it will take time for compiler optimization techniques to mature. Reference [6] reports that a carefully coded matrix multiply written for the i860 which achieved 62 MFLOPS out of the possible 66 MFLOPS peak. It is not likely that this will be easily attainable through compilers.

We are not suggesting that all microprocessors will look like the i860, but rather that they will need to explore similar design alternatives to achieve the most competitive floating point performance in their designs. Not all microprocessor manufacturers are bent on capturing peak floating point performance, realizing that workstations do so much more than solve matrix problems. SPARC and MIPS are examples. Those looking for higher floating point performance are similar in broad sense to what we have described with the i860 example. The Motorola 68000 looks very much like a CDC 7600 in some broad sense. The IBM RS-6000, a very elegant design, employs vector pipeline techniques based on a register interface. The elegance of its design comes from the fact that much of the data flow management is done within the chip and not the responsibility of (compiler generated or hand coded) assembler instructions.

Currently, new generations of microprocessors in any given line are turning over about every 18 months. By the end of the decade industry analysts predict that the power of microprocessors will be equivalent to today's latest generation supercomputers with clock speeds under 5 nanoseconds and 100s of millions of devices per chip. One cannot ignore these powerful computers on a chip as a basis for supercomputers. As the number of devices per chip grow, the architectural possibilities increase dramatically. Table 9 contrasts current generation RISC architecture from next the generation in terms of percentage of device utilization.

Table 9 RISC Microprocessor on-Chip Device Trends

	Current Gen.	Next Gen.
Floating point	20	15
Cache/Memory	24	40
Human Interface	15	20
Communication	15	9
CPU(s)	15	12
Graphics	6	10
On-board Testing	5	5
Hardware Scheduling	0	9

What Table 9 does not indicate is that new generation microprocessor chips will likely have up to 8 CPUs, and that hardware scheduling will be required for on-chip intraprocessor communication as well as communication to off-chip devices, such as microprocessors on other chips. This latter feature will promote massively parallel systems. The one clear trend is for increased memory size. This memory will be divided among local caches for each CPU and a main memory whose size will rival the memories of today's supercomputers. This increased real estate for memory will open the opportunity for increased floating point performance and may require more RISC designers to use vector instructions as discussed in the i860 example above. Current (non-vector) RISC architectures are already employing overlapped instructions, pipelined (or phased) instruction processing, as well as parallel instruction processing to achieve greater performance. As these techniques become inadequate to alleviate the overhead of instruction processing, vector instructions become the logical first choice, and multiple CPUs follow close behind [This sounds very similar to the history of supercomputer design reviewed earlier].

The complexity of microprocessor programming will far exceed current supercomputer compiler complexity owing to the large number of new tasks such as scheduling, testing, graphics and human interface tasks. However, the technology of writing these compilers will no doubt have its historical origins in the vector compilers and vectorizing software developed over the last decade for supercomputers.

The challenge of designing supercomputers in future years will likely be one of harnessing the power of a massive number of these supercomputers on a chip, and major design decisions will be those dealing with system cooling, packaging, communication topology and programming paradigms for large arrays of microprocessors within a single system. The most difficult area will be the programming paradigm and decisions about code development tools and methods. The software challenge will be to create from our stock of important applications a number of new applications that can exploit massive parallelism. In aerodynamics new applications of interest include optimal wing design (the inverse problem of CFD), aeroelastic control and other multiple discipline problems involving CFD. (New applications based on changing analysis programs into design programs are among the most promising uses of new parallel computers. See Ref. [5].)

Regardless of the application area, massively parallel systems can only be effectively used by developing computational approaches that can "scale" with the increasing numbers of CPUs. This is more complex that developing an effective parallel code for an 8-CPUs shared memory computer such as a Cray Y-MP. When writing parallel code there is generally overhead for synchronization of independent tasks. This is much easier to control for relatively small numbers of CPUs. A scalable algorithm will not have this overhead grow more than linearly with increases in the degree of parallelism. Very few production applications have been developed with this property. This will be the most major barrier to the market acceptance of future systems. Nevertheless, the author recalls similar statements about vector processing 15 years ago, and now it is a common and well understood computational complexity. The key difference in the massively parallel challenge is that existing programs are not going to "migrate" via some magic compiler technology. There is no automatic approach to parallelization of this magnitude. History has shown that, even when it is possible to create an automatic solution such as a good vectorizing compiler, the tool takes years to perfect. We will see better analysis and debugging tools for parallel processing, but they will be most effectively used in new program developments as opposed to migration of existing software. This will also cause application benchmarks as a tool for computer selection somewhat ineffectual.

8. CONCLUSION

Looking at the current crop of supercomputers, we find that all the players have made reasonable decisions and comprises against their own constraints. It is clear that the "balance" of the Y-MP's single CPU architecture, and the tools to efficiently utilize its multiple CPUs, allow it to compete with hardware boasting much higher peak performance rates on traditional problems. It is equally clear that the second generation of Japanese machines (reviewed here) will be much more competitive than their first offerings. While their latency to memory will continue to drive their respective $N_{1/2}$ values higher, the introduction of multiple CPUs, overlapped instructions, larger memories, and aggressive compiler tools will provide them a technological base from which to compete, particularly on very large easily vectorizable problems. These machines will not run as close to their peak performance potential as the Y-MP, but will run faster on some problems owing to the sheer advantage of raw power and low cycle times.

Over the longer term, all the high-end vector architectures will receive serious challenge from the more massively parallel machines, particularly when competing in special problem classes. There are a number of specialized problems that are amenable to highly parallel computing algorithms, and for those applications there could be a large price/performance advantage. Servicing single application environments, or as nodes on an integrated system, massively parallel machines may gain market acceptance. The decade of the 1990s will, no doubt, be challenging for users and manufacturers alike.

9. REFERENCES

[1] Gentzsch, W. and Neves, K. W., "Computational Fluid Dynamics: Algorithms and Supercomputers," ed. H. Yoshihara, NATO AGARDograph No. 311, March 1988, pp. 182.

[2] Hockney, R and Jesshope, C., "Parallel Computers 2," Adam Hilger LTD, Bristol, 1988

[3] Neves, K., "Mathematical Libraries for Vector Computers," Computer Physics Communications, Vol. 26, 1982, pp. 303-310.

[4] Dongarra, J., Lecture Notes, "Supercomputer Seminar," San Diego State University AMCEE Education Seminar, KMPS-TV, March 4, 1987.

[5] Kowalik, J. and Neves, N., "Supercomputing: Issues and Challenges," Supercomputing, NATO ASI Series F : Vol. 62, ed. J. Kowalik, Springer Verlag, June, 1989, pp. 3-43.

[6] Fried, S., "Personal Supercomputing with the Intel i860," BYTE, Jan., 1991, pp. 347-52.

VI NAVIER-STOKES BENCHMARK TESTS

Kozo Fujii*
*** Associate Professor,**
The Institute of Space and Astronautical Science
3-1-1, Yoshinodai, Sagamihara, Kanagawa, 229 JAPAN

and

Hideo Yoshihara†
† Former, Liaison Scientist, Office of Naval Research
Roppongi, Tokyo, 106 JAPAN

Current address: 8620, NE, 20th St. Bellevue, WA, 98004, U.S.A.

SUMMARY

Results of recent benchmark tests for current supercomputers on the unsteady Reynolds-averaged Navier-Stokes simulation code are given. All the top-class supercomputers are benchmarked and the results are discussed from CFD view point. The results indicate that peak performance is only one indication of supercomputer performance. Compared to the GFLOPS rate that have been multiplied by increasing the number of pipelines and processors, data transfer rate from the memory to vector unit is not easily improved. For practical CFD problems, data supply to vector unit as well as compiler improvement to utilize number of multi-pipelines and multi-processors efficiently are important as well as peak performance. Responsibility of compilers, although the improvement is not as dramatic as hardwares, is becoming more and more increased.

Based on the benchmark results, capability of the current supercomputers are evaluated and the requirement on future supercomputers that makes the CFD simulation feasible in practical and basic fluid dynamic research applications are clarified.

1. INTRODUCTION

Although development and improvement of the solution methods for the Navier-

Stokes equations is, without doubt, one of the important factors that promoted the spread of Computational Fluid Dynamics (CFD) research, a key important factor was obviously the development of supercomputers. Today's fluid dynamic research cannot be discussed without supercomputers. Future development of supercomputers may have to take the requirement by the CFD into account since the CFD is one of the important research fields that utilize supercomputers extensively. There are many benchmark reports available today. However, they mostly use so called kernels to evaluate the supercomputer performance. For the fluid dynamic researchers, benchmark test on real fluid dynamic codes may be useful.

With such background, benchmark test of real Navier-Stokes computer code, LANS3D-UP on top-class supercomputers was carried out by the present authors in 1989. Four supercomputers were benchmarked: CRAY Y-MP/832, Fujitsu VP-400E, Hitachi S-820/80 and NEC SX-2A. The calculations were carried out by computer company personnel over the 3-months period from 15 October to 13 January 1989. The report itself is available as Refs. [1] and [2].

Several top-class supercomputers have appeared since then. They were NEC SX-3 and Fujitsu VP2600. CRAY has also announced its new machine Y-MP C-90, originally called C-90. Hitachi did not announce its newer model, but modified the memory chips of S-820/80, and the performance was improved. Considering these changes on supercomputer technologies, all the supercomputer companies were again requested to carry out the same benchmark test late in 1990.

In this chapter, all the benchmark test results are presented with the emphasis on recent supercomputers. The results are discussed from the CFD view point, and the future requirement is clarified. Some of the issues associated with the existing benchmark test are also addressed. Benchmark test conducted recently on CRAY Y-MP C-90 is also presented (see section 6) although this was done after the benchmark period and is not included in the same tables so that the comparison becomes fair to all the companies.

2. BENCHMARK TEST FEATURES

The benchmark procedure selected was the Fujii/Obayashi code[3], one of the representatives of the three-dimensional compressible Navier-Stokes codes in use in the United States, Japan and Europe. The code is called LANS3D-UP which was a modified version of original LANS3D. The benchmark problem was for the subsonic, high-angle of attack flow over a strake-delta wing considered earlier[4].

The flow condition was set as: Mach number 0.3, an angle of attack 12°, and the Reynolds number based on the root chord 1.3×10^6. In the original computation[4], laminar flow was assumed. In both of the present benchmark tests, on the other hand, a subroutine for computing the Baldwin/Lomax turbulence model was intentionally added to test the ability of the compiler to handle this subroutine that is difficult to vectorize.

The mesh dimensions supplied were 119 points in the chordwise, 101 points on the half-circumference, and 71 points in the radial direction for a total of about 850000 mesh points. This mesh dimension was selected because it requires almost 220 Mbytes memory which was close to the maximum memory size in double precision (64 bits computations) for some of the machines to be benchmarked. Three cases were calculated using this mesh size as follows.

1. Calculations with the code as furnished using the standard compilers. For the CRAY Y-MP/832 calculations with both one and eight CPU's using the autotasker in the latter. Note that autotasker seeks to distribute automatically and optimally the workload to the CPU's at the subroutine to DO-loop levels (200 iterations were requested).

2. Calculations using the code as furnished, but permitting additional Fortran-compatible compiler tuning including use of compiler directives (200 iterations were requested).

3. Calculations same as Case 2, but modifications of the algorithm and changes in the programming additionally permitted. (200 and 2000 iterations were requested, latter to measure the convergence).

Benchmark participants were as follows.

* **Cray Research Inc.**
  Mr. K. P. Misegades, Manager, CFD Application Group, Cray Research Inc., Mendota Heights, MN 55120, U.S.A.,
  and
  Dr. T. Saito, Cray Research Japan, Inc., 6-4, Ichibancho, Chiyoda-Ku, Tokyo, 102, Japan.

* **Fujitsu Ltd.**
Mr. N. Tahara, Manager, Scientific Application Section, Fujitsu Ltd., 1-17-25, Shinkamata, Ota-Ku, Tokyo, 144, Japan.

* **Hitachi Ltd.**
Mr. S. Kawabe, Senior Engineer (currently Manager),
and
Mr. K. Ishii, Computer Development Dept., Hitachi Ltd., 1 Horiyamashita, Hadano, Kanagawa, 259-13, Japan.

* **NEC Corp.**
Mr. S. Mineo, Supervisor, 3rd Systems Support Department, NEC Corp., 2-7-17, Shiba, Minato-Ku, Tokyo, 105, Japan (for SX-2A),
and
Mr. Y. Kobayashi, Supervisor, Supercomputers Marketing Promotion Division, NEC Corp., 5-7-1, Shiba, Minato-Ku, Tokyo, 108-01, Japan.

Table 1 shows the hardware feature of the supercomputers. The table includes multi-processor machine SX-3/44 which has not been available at the time of the benchmark tests, and CRAY Y-MP C-90 which is to be delivered in 1992. Note that Y-MP C-90 adopted the architecture of multi sets of pipelines and NEC now has multi-processor machine.

Although VP2600/10 (10 indicates the machine having one scalar unit and one vector unit) was used for the benchmark test, the table includes VP2400/40 because the feature of this machine is noteworthy. Fujitsu attached importance to the fact that even 90 % vectorized code makes the vector unit busy less than 50 % of the total CPU time. Besides, many of the supercomputer users are not necessarily well acquainted with the vectorization techniques, and thus vectorization ratio stays, in average, 90 % to 95 %. If so, adding one more scalar unit may be useful. In other words, two scalar units share one vector unit instead of one vector unit with one scalar unit. They are called VP2000/20 series. For instance, VP2600/20 has two scalar units and one vector unit. In the case of VP2400, VP2400/10 has one scalar and one vector units and VP2400/20 has two scalar units and one vector unit. VP2400/40 has an special extension and has two vector units and four scalar units. Each vector unit has 2.5 GFLOPS (note: clock cycle is not the same as 2400/10 and /20) performance and in total, machine performance becomes 5 GFLOPS. This machine has versatile feature depending on the user's need. One job can use two vector units (multi tasking) or each vector unit can be used by a single job. Two scalar units can be shared and used with one vector unit, or only one unit is used with one vector unit, and the additional scalar unit is used for one scalar job. Fujitsu seems to improve the 'system performance' rather than just achieving peak performance represented by the GFLOPS rate. Although this machine was not benchmarked, the philosophy of this machine is important to note. Note that there is no VP2600/40.

Further details of all the newest models are described in the other chapters by the representative person of each supercomputer manufacturer. Here, only a summary is presented for reader's convenience.

3. BENCHMARK TEST RESULT – 1

First, benchmark test result in 1989 is presented. Four supercomputers were benchmarked within Table 1. They are: CRAY Y-MP/832, Fujitsu VP-400E, Hitachi S-820/80 and NEC SX-2A.

The benchmark result is summarized in the Table 2. Note that only the prototype machine was available for CRAY Y-MP at the time of the present benchmark test, and the clock cycle was 6.41 nsec instead of 6.0 nsec. One of our interest is the computational speed actually attained that is a measure of performance. It is obtained by dividing the total operation count of the calculation by the CPU time. The operation count of the original code was obtained in Case 1 from the measured performance of the single-processor CRAY Y-MP where an actual speed of 0.175 GFLOPS was achieved with a CPU time of 3454 seconds.

Based on this operation count, the actual speeds of the supercomputers were computed. The result is also shown in Table 2. The peak and sustained GFLOPS rates are graphically plotted in Fig. 1 for easier understanding of the result. The actual GFLOPS rates indicate how fast these supercomputers are for the realistic fluid dynamic program code. Except CRAY Y-MP that is a multiple-processor machine, the speed obtained is about 500 MFLOPS. This number clearly indicates how important supercomputers are for the computational fluid dynamics research (Note that the performance of general-purpose computers is at most a few ten's MFLOPS). The actual/peak ratio is another interesting number that shows how efficient was the computation carried out. Actual speed of 30 % of the peak is a good number for an implicit Navier-Stokes code like this, and the result indicates that all the compilers are sufficiently smart for vectorizing most of the heavy DO-loops.

The result that the sustained/peak ratio for all the Japanese supercomputers stays about 30 % is probably due to the fact that increasing the number of pipelines decreases the effective vector length for each pipeline and thus the performance is not linearly improved with the increase of pipelines. Reducing the overhead for vector operations may be necessary to improve this. Total performance is, however still very much improved by the concept of multi-pipelines as can be noticed from the actual GFLOPS result. Another factor may be the data transfer capability from the main memory to vector processor and the memory access speed (see Table 1). These may be important for fluid dynamic codes as large-scale computations are under current interest. The sustained/peak ratio that CRAY Y-MP shows is a remarkable number and proves that this machine is well balanced for vector processing, data transfer, memory access etc. with respect to the present problem. One thing to note is that Fujitsu VP400E and Hitachi S820/80 adopt hardware-chained add/multiply pipelines. Since these pipelines are not efficiently used at least for this benchmark code, the peak performance could be reduced. Sustained/peak ratio, then becomes better as indicated in the parenthesis in Table 1 when the chained add/multiply pipe is counted as one pipe.

It is important to keep the computer program tidy so that it is easily read and modified. The reason is that the current CFD technology is not in the matured stage that anyone can use them with no expertise and some part of the program has to be modified depending on the application problems. Too excessive effort for achieving high performance is, in the authors' opinion, beyond the realm of CFD. The computer code used in this benchmark was written under that philosophy and the program was vectorized only by a basic strategy that nothing that would make the program messy was adopted. So, the typical dimension for DO-loops is at most 119. That makes the vector length short and that caused relatively low actual/peak ratio on Japanese supercomputers. It would be possible to make vector DO-loops longer by expanding two-dimensional array into one-dimensional array to improve the performance (and this is actually what NEC did to improve the performance in Case 3), but such should be taken care of by the compilers. There remains possibility to improve the performance by the development of smarter compilers that handle this.

Both hardware and compiler should be well developed for good supercomputers that are easy to use. Besides, one of the advantages of the CFD over experimental approach is portability which means we can bring and use the code anywhere in the world so far as computers are available. In that sense, the program should run on most of the general-use computers as well as supercomputers. All the Japanese supercomputers, benchmarked here, are single-processor machines and the users can enjoy good performance with doing nothing except basic vectorizing effort. This effort is relatively simple and what we have to do is to remember several important facts about supercomputers. On the other hand, Cray supercomputers are multi-processor machines and thus have required the special effort for the macrotasking or microtasking to get good performance. The 'autotasker' with FPP (Fortran pre-processor) removed this problem and now users can use the machines without too much effort. In other words, the compiler basically takes care of parallelization. Although there still needs for additional improvement (for instance, in this benchmark four directives are put into the code to help the compiler), the development of this autotasker has a strong impact on this benchmark result. So far as the present code is concerned, good performance is obtained with little modification to help the autotasking compiler on CRAY Y-MP.

4. BENCHMARK TEST RESULT – 2

The second benchmark test was conducted from late in 1990 to early in 1991. As an exception, Hitachi data were obtained in summer of 1990. CRAY Y-MP/832 data was updated since clock cycle is now 6.0 nsec and compiler may have been improved. Benchmark test by Fujitsu VP2600 was conducted using the machine located at the National Aerospace Laboratory in Chofu, Tokyo.

The benchmark result is summarized in Table 3. Again, the peak and sustained GFLOPS rates are graphically plotted in Fig. 2. The actual GFLOPS speeds were again computed based on the operation count obtained on the CRAY Y-MP in 1989.

First is to note that all benchmarked machines almost realized the requisite GFLOPS performance needed in a practical CFD problem as suggested in the following. In the 7th International Conference on Numerical Methods in Fluid Dynamics held at Stanford university and NASA Ames Research Center in June, 1980, Prof. Dean Chapman at Stanford University gave an assessment of CFD technologies[5]. One of the tables he showed is reproduced as Table 4. He estimated that Reynolds-averaged Navier-Stokes computation that simulates the wind-tunnel experiment for wing-body configuration would require 2 million grid points. That would need 60 million words (480 Mbytes for 64 bits computation) computer memory and the computer speed of 600 MFLOPS to give a result within one hour of computer time. He also estimated the simulation under real flight condition would require 4 million grid points that correspond to 100 million words (800 Mbytes). The required speed for one-hour simulation was 1.2 GFLOPS. Current supercomputers almost satisfy the requirement of the memory size and the computer speed that Chapman expected about 10 years ago.

NAS (Numerical Aerodynamic Simulation) program at NASA was completed in 1987[6]. This important project was initiated so that the United States takes a leadership in the area of Computational Fluid Dynamics in the future. The requirement for the HSP-1 (High speed processor 1) in the IOC (Initial Operating Configuration) was 250 MFLOPS for the real performance on the CFD applications. The requirement for the HSP-2 that was planned to be installed 2 years later was 1 GFLOPS speed. The performance of about 1 GFLOPS was one of the criterion that should be achieved, and it is now realized. Measurement of actual speed of more than 1 GFLOPS indicates that these criteria have been broken through.

Second to note is that sustained speed does not necessarily reflect the peak performance. In other words, strong variation exists on the sustained/peak performance ratio. Which machine shows the best result is not important. The point is that maximum or peak performance only reflects the processor capability but on the other hand, memory access speed, data supply to the processor and other factors are becoming more and more important for the performance on real problems especially in CFD applications. This is being pronounced as supercomputer performance is improved. For instance, Hitachi S820 shows the much better result in the recent benchmark simply because the memory chip is improved.

The performance of the original code (Case 1) may also be important. The result in 1989 is graphically represented in Fig. 3 and the result in 1991 is shown in Fig. 4. Since no good result was obtained without 'directives' on multi-processors of CRAY Y-MP, the data for a single-processor is plotted for a fair comparison. Let's compare

Figs. 1 and 3. On all the supercomputers, the performance of Case 1 (sustained speed in Fig. 1) and Case 3 (sustained speed in Fig. 3) showed little difference in the benchmark of 1/89. This indicates that the compilers for supercomputers those days were well in the matured stage. In the benchmark test of 3/91, Fujitsu and NEC show remarkable improvement by their effort to obtain high performance in Case 3. Both companies had almost the same strategy to improve the performance. Namely, double and triple DO-loops are expanded to achieve much longer vector length. Although the result for the original code shows more than 1 GFLOPS on these machines and the peak/sustained ratio is about 0.2 to 0.3, (which is not bad), their compilers still have room to be improved. This is especially true for the NEC SX-3. It seems that compiler's responsibility becomes too heavy for newer multi-pipelined machines going for high peak speed.

Single-processor performance can be recognized in this graph. There are not many users that utilize multi-processors. CRAY Y-MP multi-processors are conventionally used by many people at the same time, and in that case, single-processor is allotted to each user. Single processor performance may be important for considering this situation (Note that throughput is much different). Although CRAY Y-MP single-processor performance is lower than other supercomputers because of its fewer pipes/CPU, the result proves its wonderful feature in the sustained/peak ratio. The fact that the sustained/peak ratio on this typical CFD code exceeds 50 % is remarkable. Therefore, very good throughput is expected when the machine is shared by number of jobs using relatively small memory.

5. FINAL REMARKS ON BOTH BENCHMARK TESTS

5.1 Assessment of the Result

The purpose of the present benchmark was to evaluate the relative performance of current supercomputers using an implicit Navier-Stokes code on a representative fluid dynamic problem. The benchmarked computers covered a wide range of peak speeds, belonged to different generations, and differed greatly in other factors such as costs. The benchmark results must be accordingly tempered with these differences in mind. The benchmark supercomputers fell in two basic categories. The first class (Category

1) was composed of the Fujitsu, Hitachi and NEC computers where many arithmetic processing pipes were driven via a single instruction path. The second class (Category 2) was composed of the CRAY Y-MP/832 that was characterized by having several independent instruction paths.

To neutralize the wide range of peak speeds, performance was mainly assessed in terms of the sustained/peak speed ratio. Findings of the benchmark can be summarized as follows.

1) Sustained/peak speed ratios of the Category 1 computers were significantly lower than that of the Category 2 computers. The cause in large part was due to the pipeline overhead arising by the dividing the vector length across the many pipes. The high sustained/peak ratio cannot be obtained unless the vector length per pipe is sufficiently large. Despite the near-100 percent vectorization, subroutine calls and nested DO-loops can significantly degrade the sustained/peak performance ratio. Under certain circumstances, compilers take care of in-line expansions and this problem is partially solved by the compiler improvement. However, still more compiler improvement is required. Category 2 computers, although require small user's effort such as inserting compiler directives, show good result for this benchmark code that have many double and triple DO-loops that are typical for many CFD codes.

2) Peak performance is only one aspect of supercomputer performance. In many large-scale CFD simulations, data supply from memory to vector register are critically important. Recent technologies employing increasing number of pipes and processors have multiplied the peak performance of the machines. Memory latency and data transfer rate from memory to vector register must keep pace with the peak speed for the users to enjoy the high performance on the real CFD programs.

Japanese computers benchmarked here fall into Category 1 and only Cray computers fall into Category 2. However, basic hardware concept seems to move in the same direction. Cray's new machine, Y-MP C-90 utilizes the idea of multi sets of pipelines to improve the performance. NEC SX-3/44 is a multi-processor machine like CRAY Y-MP although its benchmark data has not been available yet. Fujitsu has VP2400/40 that has two vector processors. Multi-tasking compilers are now under development and will be released soon. Again, compiler capability is the key factor to achieve high actual performance for any machines. Hitachi, although it shows the worst result in this benchmark result, will soon announce its newer model which is believed to be composed of 4 processors with 8 GFLOPS performance on each processor. Benchmark test on this machine will probably show much better result. In addition, all the companies are now moving to the development of massively parallel machines. Compilers, (probably with the user interface to help users' efficient parallelization of their programs) will be much more important for the practical machine capability.

Although sustained/peak ratio was used as a measure of the machine performance to neutralize the comparison, this type of benchmark data still reflects only one particular feature of supercomputers. The following statement by Miyoshi at the National Aerospace Laboratory in Japan is important when discussing the present (and probably any) benchmark result. He says,

"Any of the current supercomputers may show different speed for the code developed on the same algorithm and applied to the same problem. The speed depends on the person who programs the computer code."

The tables and graphs shown above should not be used for a simple comparison of the performance of each supercomputer. In many benchmark tests, the result depends on how much effort the benchmark participants make. The result depends on the applications and even in one application, the result depends on the program itself as mentioned earlier.

We have to consider the overall cost, throughput capability, operating system, application softwares and else. Performance based on single-processor may be important for some users. Cost performance may be important for another users. Electric consumption may be critical. There are many factors to evaluate supercomputers. Peak performance is only one factor for the supercomputer performance. About the operating system, all the supercomputers are going to support UNIX operating system. NEC emphasizes the Super-UX which is an extended version of UNIX operating system. Fujitsu, although still keeps its original operating system named MSP, developed AVM (advanced virtual environment) where both UXP(Fujitsu's UNIX O.S.) users and MSP users work on their choice of operating system, and is extending it to AVM/EX where job scheduling becomes much more flexible. Networking is another important factor because flow visualization using the graphic workstations as post process of the flow computations is becoming more and more important.

5.2 CFD View Point

The followings are some of the conclusions of the benchmark test from the CFD view point.

We, researchers in the computational fluid dynamics are interested in how much time is necessary for a single Navier-Stokes simulation. Table 5 shows the computer time per grid per iteration, and estimated total computer time based on that number. It is crudely assumed that 2000 iterations would be necessary to obtain a converged steady-state solution.

Note that this is applicable only to steady-state solutions. Much more computer time would be required for unsteady flow simulations which are of current interest by many CFD researchers. Also to be noted is that the number here is applicable only when single module of grid can cover the flow field to be simulated. Recent research[8-10] indicates that domain decomposition is inevitable for complex flow simulations under the structured grid concept, and thus certain overhead should be considered for the communication of each domain. Strong tendency to adopt unstructured grid concept exists now to get rid of this communication problem. The efficiency of the unstructured grid approach on supercomputers compared to the structured grid approach should be an important research topic in the near future.

On any supercomputers, the CPU time required for the implicit operator that integrate time was order of 30 % of the total CPU time. Implicit time integration at least of the LU-ADI algorithm requires only 50 % more computer time compared to the explicit time integration for each iteration. Since explicit methods have severe restriction of the time step and computer time for each iteration is not much different between explicit and implicit methods, implicit time integration is useful especially for the high-Reynolds number viscous flow simulations. Thus, the main problems associated with the implicit time integration are the requirement of large memory for the matrix inversions and the short vector length rather than the required computer time.

6. CRAY Y-MP C-90 BENCHMARK REPORT

Only the simulation data based on single-processor was available for CRAY Y-MP C-90 when this chapter was written. Fortunately, physical benchmark test was conducted in time and the data became available on January 1992 before finalizing the book. The result is independently discussed since the result was obtained much later than the benchmark period and it is not fair to include such data in the same tables. The benchmark result was, however, very impressive and thus is summarized in this chapter.

The machine used for the benchmark was the one in their factory for the final testing, and thus some of the feature is slightly different from the C-90 in its catalogue. First, machine clock is 4.2 nsec instead of 4 nsec. Second, only 15 processors were available for the benchmarking. Peak performance for each CPU is 952 MFLOPS. Cray Research says the machines capability is roughly 10 % less than the final Y-MP C-90 product. The benchmark result is summarized in Table 6. Also included is the Y-MP 8 result shown in Table 3. The sustained MFLOPS and the ratio of sustained/peak is excellent on this machine although there may be small error because the sustained MFLOPS is not the directly measured data but the calculated data based on the MFLOPS rate of Y-MP/832 in Table 2 and the ratio of the present and single CPU performance of the Y-MP/832. To utilize multi-processors efficiently, four directives are put into the code in Case 3 to help the compiler that outer DO-loops can be parallelized just as the previous two benchmark tests. Use of 15 processors reduced the total elapsed time to about 1/12 of that with one processor; while in the case of the Y-MP/8, use of 8 processors cut the elapsed time to about 1/7 of that with single processors. This would suggest an increased parallelization overhead in the case of the C-90 due to the decreased granularity.

7. FUTURE REQUIREMENTS

Current supercomputer capability was evaluated based on these Navier-Stokes benchmark tests. Required computing time and memory size was roughly estimated based on the benchmark result. The result is summarized in Table 7. Four types of typical computations are picked up; 2-D steady, 3-D steady, 2-D unsteady and 3-D unsteady. Computational grid points are assumed based on the authors' experience. Since number of the computational grid pretty much differs depending on the geometry complexity, two cases are considered; simple geometry and complex geometry. Sustained speed of supercomputers is roughly estimated to be 1 to 2 GFLOPS. "n" in the table can be small or large and different for each column. Two dimensional Navier-Stokes equations for simple geometry could be solved within a minutes or so. It means that simulation of transonic viscous flow over an airfoil could be done within about a minute, and we could obtain $C_L - \alpha$ curve, drag polar curve (if drag is accurately computed) and some more curves within one hour. The turn-around time would be satisfactory to the design process.

Dr. Paul Rubbert at Boeing Commercial Airplane Group says in his paper[7] "The goal of future developments in CFD must encompass more than merely solving the right physics about the right geometrical shape. It must include also a strong requirement for doing so in a timely manner so as to produce a significant reduction in aerodynamic design flow time. Recent experience of CFD on the Boeing 777 program indicates that timeliness can be defined as no less than one problem turn-around in a 24-hours period, with several turnarounds in that time frame being preferable". This required time frame includes all the processes necessary for the CFD simulation such as grid generation to initialize the simulation and the post process to visualize the simulated result. Three-dimensional steady flow simulation, which would require about tens of computer time for complex configuration would not still satisfy this timeliness requirement. Besides, some people say Navier-Stokes simulations are necessary only for complex configuration and they are not necessary for simple configuration that could be well simulated with nonlinear potential equations with boundary layer corrections. The computer time necessary for the steady-state simulation of three-dimensional complex configuration should be reduced to, say, less than one hour.

Airplane design is of course one of the important applications of CFD but there are many more application fields. The requirement for the computer time would be different for each application field. For instance, CFD can be a tool for understanding fluid physics. The computer time that satisfies timeliness could be longer for these problems.

Unsteady flow simulations may be more necessary than steady flow simulations. They generally require more grid points. The computational results could be obtained with similar grid size as steady flow simulations but they usually lack grid resolution because in many cases, separation vortices, shear layers emanate from the body surface and the resolution not only near the body but also away from the body becomes critically important for reliable simulations. In the case of wave propagation problems such as blast wave simulations, shock waves, contact discontinuities and other waves move in time and, thus, grid resolution in whole region to be simulated is required. In addition, time steps to be integrated become much smaller than steady flow simulations because time accuracy should be maintained. Currently several tens of hours computer time is necessary even for the simulation over simple body geometry. Flow simulations for complex configuration would require the grid points that could not fit into main memory size.

Estimation of required computer capability to make these simulations feasible is summarized in Table 8. Here, only three-dimensional simulations are considered because two-dimensional simulations are not critical. Since steady flow simulations are to be used as a design tool, even complex geometry should be treated within one hour or so, so that several results are obtained in 24 hours including pre and post process. For physically important flow simulations, the point is if it is possible or not. A hundred hours may be still satisfactory for the simulation that gives understanding of flow physics from research point of view.

Now, Table 9 shows comparison of the current computer power and the required computer power that satisfies these requirements. This result, by accident, coincides with the Mr. Miyoshi's estimation at the National Aerospace Laboratory based on industrial requirements. The supercomputer architecture that enables this severe requirement is intensively discussed in Chapter V.

8. FINAL REMARKS

In this report, global supercomputer performance is discussed for the computation of the Navier-Stokes code. Characteristics of each machine would become clearer if we picked up some of the important DO-loops and measure the speed for the computation of those particular DO-loops. This may give much more information to the users of each supercomputer (although it may also bring weak points into relief). Currently, Mr. Hajime Miyoshi at the National Aerospace Laboratory assembles a few hundreds of typical DO-loops for many CFD codes and started the benchmark test under this idea. Some of the DO-loops in the LANS3D-UP will be included in his benchmark test. This may give us more information and will be helpful to modify our codes to be adequate for each machine if necessary.

9. ACKNOWLEDGMENT

The authors would like to express their appreciation to the benchmark participants listed earlier and to the managements of Cray Research Inc., Fujitsu Corp., Hitachi Corp. and NEC Corp. for making benchmark tests possible. Much of the material contained here was extracted from reports submitted by the benchmark participants.

10. REFERENCES

[1] Fujii, K. and Yoshihara, H., "A Navier-Stokes Benchmark for Japanese and U.S. Supercomputers," ONRFE SCI INFO BUL 14(2)89, 1989.

[2] Fujii, K., "Capability of Current Supercomputers for the Computational Fluid Dynamics," Supercomputing '89, Reno, Nevada, Nov., 1989.

[3] Fujii, K. and Obayashi, S., "High-Resolution Upwind-Scheme for Vortical Flow Simulations," J. Aircraft, Vol. 26, No. 12, December, 1989, pp. 1123-1130.

[4] Fujii, K. and Schiff, L. B., "Numerical Simulation of Vortical Flows over a Strake-Delta Wing," AIAA Journal, Vol. 27, No. 9, 1989, pp. 1153-1161.

[5] Chapman, D. R., "Trends and Pacing Items in Computational Aerodynamics," *Proc. 7th International Conference on Numerical Methods in Fluid Dynamics*, Lecture Notes in Physics 141, Springier-Verlag, 1980.

[6] Bailey, F. R., "NAS-Current status and Future Plans," Supercomputing in Aerospace, NASA CP2454, 1987.

[7] Rubbert, P., "On the Continuing Evolution of CFD for Airplane Design," Conference Textbook, Supercomputing Japan-91, April, 1991.

[8] Buning, P. G. et al., "Numerical Simulation of the Integrated Space-Shuttle Vehicle in Ascent," AIAA Paper 88-4359, 1988.

[9] Rizk, Y. and Schiff, L. B., "Numerical Simulation of the Viscous Flow Around a Simplified F/A-18 at High Angles of Attack," AIAA Paper 90-2999, 1990.

[10] Fujii, K., "A Zonal Method for Practical Fluid Dynamics Problems," Proc. ISCFD-Davis, Sept., 1991.

Table 1 Supercomputers and Their Hardware Characteristics

Computer	Peak Peformance (GFLOP)	Clock (nsec)	Memory Size (Mbytes)	Mem/Reg Bandwidth (Gbytes/sec)	No. Proc.	1st Delivery
CRAY Y-MP/832	2.67	6.0(6.4)	1024	42.4	8	March, 1988
Y-MP C90	16.0	4.0	4096	256	16	1st Qt., 1992
Fujitsu VP-400E	1.7	7.0	1024	4.56	1	Dec., 1987
VP2600/10	5.0	3.2	2048	20	1	June, 1990
VP2400/40	5.0	3.2	2048	20	2+2su	
Hitachi S-820/80	3.0	4.0	512	16	1	Dec., 1987
NEC SX-2A	1.3	6.0	1024	11	1	June, 1988
SX-3/14	5.5	2.9	2048	22	1	
SX-3/44	22.0	2.9	2048	80	4	

Table 2 Benchmark in January, 1989

Computer	CPU (sec)	Elapsed (sec)	Sustained (GFLOPS)	Peak (GFLOPS)	Ratio	Clock (nsec)
CRAY Y-MP/832						
1. (1 CPU)	3454.0	3455.0	0.175	0.313	0.56	6.4
2. (8 CPU)	3313.0	478.0				
3. (8 CPU)	3304.0	489.0	1.23	2.50	0.49	
(2000 it.)	33028.0	4675.0	1.29			
Fujitsu VP-400E						
1.	1530.42	1690.33	0.395	1.7(1.14)[b]	0.23 (0.35)[b]	7.0
(2000 it.)	15275.98	15504.66				
3. [a]	1372.20	N/A	0.440		0.26	
Hitachi S-820/80						
1.	1005.17	1068.0	0.601	3.0(2.0)	0.20(0.30)	4.0
3.	979.26	1106.0	0.617		0.21	
(2000 it.)	9718.69	98180.0				
NEC SX-2A						
1.	1459.75	1487.2	0.414	1.3	0.32	6.0
3.	1207.38	1231.0	0.502		0.39	
(2000 it.)	12048.19	12071.9				

[a] Case 3 data here was the measured data obtained later and only 200 iteration was available.

[b] Chained add/multiply pipe counted as one pipe.

Table 3 Benchmark in March, 1991

Computer	CPU (sec)	Elapsed (sec)	Sustained (GFLOPS)	Peak (GFLOPS)	Ratio	Clock (nsec)
CRAY Y-MP/832						
1. (1 CPU)	2813.0	2834.0	0.215	0.334	0.64	6.0
2. (8 CPU)	3032.0	1529.0	0.399			
3. (8 CPU)	2915.0	402.0	1.52	2.67	0.57	
(2000 it.)	29043.0	3874.0				
Fujitsu VP2600/10						
1.	431.3	435.2	1.40	5.0	0.28	3.0
3.	276.5	281.7	2.19		0.44	
Hitachi S-820/80[a]						
1.	765.4	N/A	0.790	3.0(2.0)[b]	0.26(0.40)[b]	4.0
3.	723.7	N/A	0.836		0.28	
NEC SX-3/14						
1.	534.8	553.9	1.13	5.5	0.21	2.9
3.	288.2	295.3	2.10		0.38	
(2000 it.)	2875.2	2910.3				

[a] With improved memory chips.

[b] Chained add/multiply pipe counted as one pipe.

Table 4 Chapman's Assessment in 1980[5]

	Re number	Grid points	Memory Capacity (MW)	Speed for 1h Run (MFLOPS)
Wing body(W.T.)	10^7	10^6	60	600
Wing body(Flight)	10^8	$4x10^6$	100	1200

Table 5 Estimated Computer Time for Navier-Stokes Computations

Computer	Computer Time /Grid /Iteration (μsec)	One Step for Million Grid Points (sec)	Computer Time for One Million Grid Points (hours)
CRAY Y-MP/832	2.355	2.355	1.31
Y-MP C90*	0.463	0.463	0.26
Fujitsu VP-400E	8.040	8.040	4.47
VP2600/10	1.620	1.620	0.90
Hitachi S-820/80	5.738	5.738	3.19
NEC SX-2A	7.074	7.074	3.93
SX-3/14	1.688	1.688	0.94

* This data is added later in Jauary 1992.

Table 6 CRAY Y-MP C-90 Benchmark

Computer	CPU (sec)	Elapsed (sec)	Sustained (GFLOPS)	Peak (GFLOPS)	Ratio	Clock (nsec)
CRAY Y-MP C-90						
1. (1 CPU)	938.9	969.1	0.644	0.952	0.697	4.2
3. (15 CPU)	1153.6	79.0	7.90	14.28	0.553	
CRAY Y-MP/832						
1. (1 CPU)	2813.0	2834.0	0.215	0.334	0.64	6.0
3. (8 CPU)	2915.0	402.0	1.52	2.67	0.57	
(2000 it.)	29043.0	3874.0				

Table 7 Current Computer Capability for CFD (Rough Estimation)

	Simple Geometry		Complex Geometry	
	Grid Points	Computer Time	Grid Points	Computer Time
2-D Steady	$n \times 10^4$	1 ~ 5 min.	$n \times 10^5$	30 min. ~ 3 h
3-D Steady	$n \times 10^6$	1 ~ 5 h	$n \times 10^7$	10 ~ 50 h
2-D Unsteady	$n \times 10^5$	10 ~ 100 h	$n \times 10^6$	?
3-D Unsteady	$n \times 10^7$	200 h ~	$n \times 10^9$	X

Table 8 Required Computer Capability

	Simple Geometry		Complex Geometry	
3-D Steady	$n \times 10^6$	5~10 min.	$n \times 10^7$	1 h
3-D Unsteady	$n \times 10^7$	10 h	$n \times 10^8$	100h

Table 9 Required Computer Power for CFD

	Current	Required
Maximum Speed	1 ~ 5 GFLOPS	100 ~ 200 GFLOPS
Main Memory	2 GBytes	30 ~ 50 GBytes

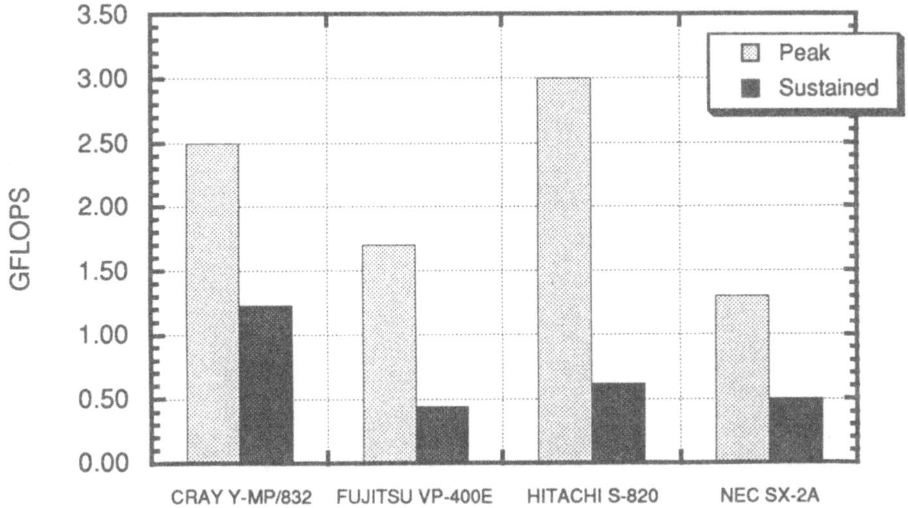

Fig. 1 Peak .vs. Sustained Performance of Supercomputers (ONR/ISAS CFD Benchmark Test Result, 1/89)

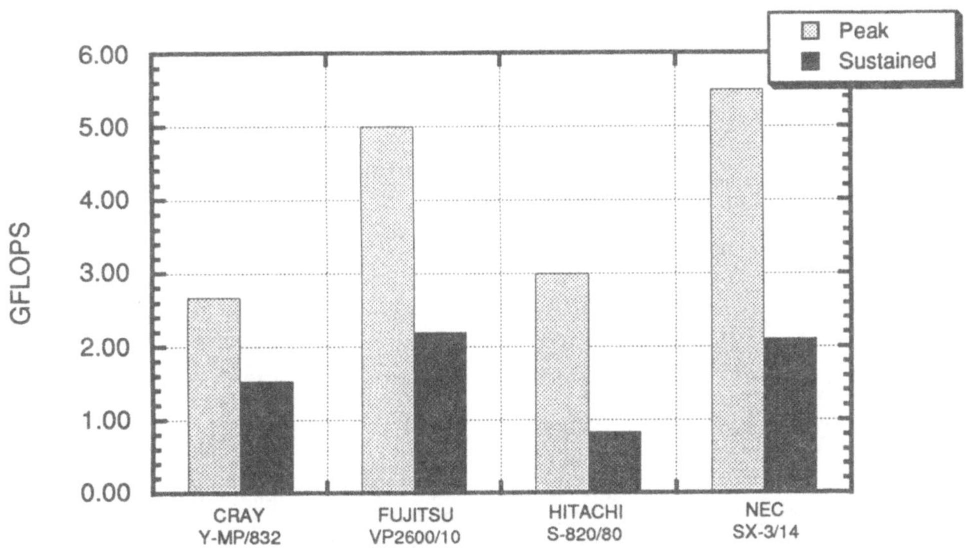

Fig. 2 Peak .vs. Sustained Performance of Supercomputers (ONR/ISAS CFD Benchmark Test Result, 3/91)

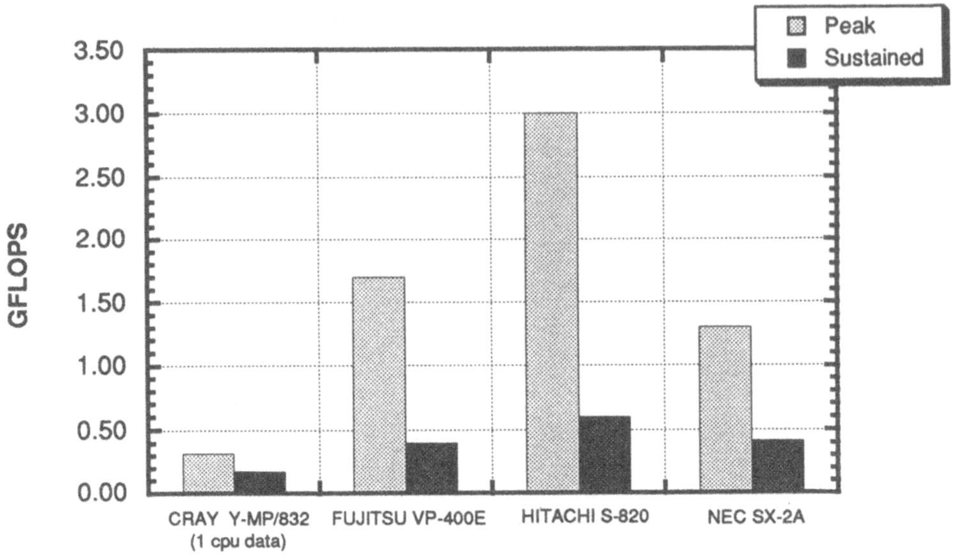

Fig. 3 Original Code Maximum/Sustained Performance (ONR/ISAS CFD Benchmark Test Result, 1/89)

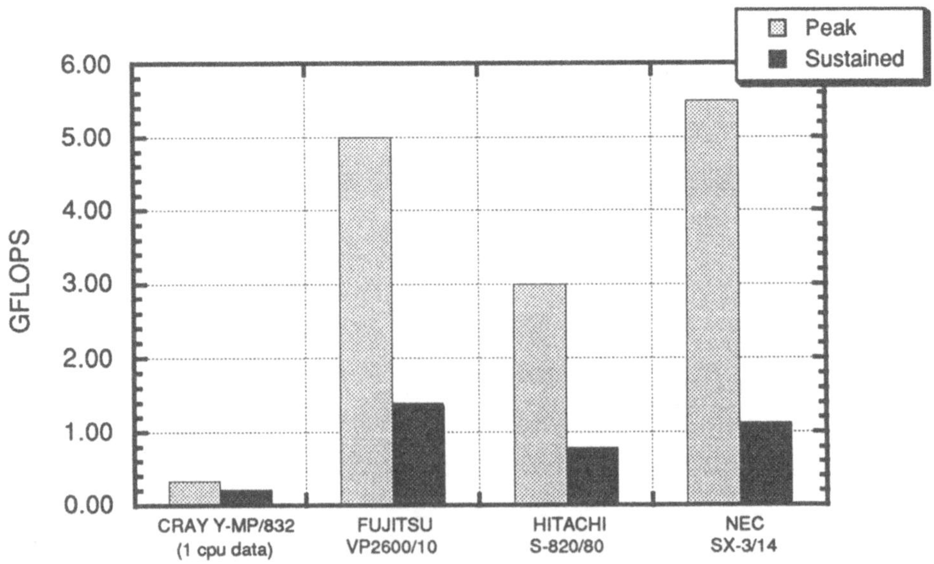

Fig. 4 Original Code Maximum/Sustained Performance (ONR/ISAS CFD Benchmark Test Result, 3/91)

VII VECTORIZATION AND PARALLELIZATION TECHNIQUES FOR MODERN SUPERCOMPUTERS

Wolfgang Gentzsch
College of Regensburg, Germany
Roentgenstr. 13, D-8402 Neutraubling, GERMANY

SUMMARY

In this article, the main aspects arising with the implementation of computer programs and numerical algorithms on vector and parallel computers are discussed with the aid of two examples from finite difference and finite element methods for the solution of partial differential equations. In addition, several results are presented for the parallel machines from ALLIANT, CONVEX, CRAY, IBM and iP-SYSTEMS.

1. INTRODUCTION

Computational Physics is becoming an increasingly powerful tool for the simulation of complex problems, e.g. in aerodynamic design of aerospace systems, as a result of both improvements in numerical algorithms and in the processing speed and storage capacity of new generations of computers. As the next generation of supercomputers becomes available much current work may be expanded to address more complex configurations, geometries and applications.

An increasing number of issues related to computer science are appearing in the development of advanced computational capabilities. With the treatment of more complex 3-D configurations, the issues of

* vectorization,
* multitasking,
* large data bases,
* analysis of computed results,
* 3-D graphic displays,

* surface and grid generations,

become increasingly important. Beside the algorithm aspects, the progress in Computational Physics will also depend on the power and storage capacity of the computers available. Therefore, an intimate knowledge of the supercomputer hardware and its impact on the algorithms is essential.

Improvement in supercomputer performance only by the progress in technology is nowadays getting more and more difficult. The last remarkable increases in performance have been made mainly because of improvements of the architecture, by vector and parallel processing. Whereas today the average performance of one vector processor is about 30-100 MFLOPS, an increase in performance by a factor of two to four is predicted because of improvements in technology and in compilers in the next few years, and another increase by a factor of 10 and more in using more than one processor for the parallel execution of one program.

For this reason we are now at the very beginning of a new epoch for the computation of engineering and scientific applications. Most of the existing algorithms and programs were written only for scalar and vector computers with one CPU. Now these algorithms have to be restructured or even new ones developed for parallel computers.

2. BASIC ASPECTS OF VECTOR AND PARALLEL PROCESSING

In this section, we briefly summarize the most important features of the vector and parallel computer architectures currently in use, the knowledge of which is absolutely necessary for a successful implementation of software and algorithms. For more details, see the other contributions in this book. In addition, the main properties of the vectorizing compilers are outlined.

2.1 Vector Architectures and Vector Processing

Most of todays supercomputers have a few very powerful processors coupled via very large shared memory, such as the CRAY Y-MP (8 processors), Fujitsu (4 scalar and 2 vector processors), and NEC SX-3 (4 processors). The vector processor module can perform multiple operations per instruction (e.g. up to 64 Adds, up to 64 Multiplies with one vector instruction) in a pipelined fashion. These pipelined functional units which can work independently of each other in parallel, are

* add pipes for vector instructions like add/subtract and vector logical,
* vector multiply pipes that execute floating-point or integer multiply and divide instructions,
* multifunction pipes that perform add, multiply, or multiply and add, depending on the needs,
* move pipes that execute compress/distribute and vector indexing instructions,
* load/store pipes that move data between memory and vector registers.

Each of these pipes may be in execution at the same time. For example, the vector processor may be simultaneously performing a vector add A+B, a vector multiplication C*D and a vector load or store.

The functional units of the vector processor module often can automatically chain operations together, e.g., the first result from the vector multiply pipe may be used immediately by the vector add pipe. Another parallel feature is often the overlapping of the scalar and the vector module so that scalar computations may be done in parallel with vector computations. A very useful feature in the new Fujitsu systems is the combination of two scalar processors with one vector processor, the latter having two sets of vector registers enabling the context switching for the two independent jobs sharing the same vector processor. For more details, see the chapter dealing with the Fujitsu 2000 series.

Vector operations are performed between vector registers or, possibly, between vector registers and scalar registers. Besides the scalar and the vector registers the important Mask registers should be mentioned with each bit position corresponding to one element of the vector register. The bits of the Mask register can be used to conditionally control the execution of vector operations. For example, a conditional vector addition adds only those pairs of elements whose corresponding bits in the Mask register are set to one. The contents of the Mask register may be generated by the execution of a vector comparison instruction.

The FORTRAN compiler performs a complete analysis of a FORTRAN program's branching structure at the intermediate language level to determine those regions where vector operations are possible. This analysis in combination with a sequence of vectorization algorithms produces a transformation of the original FORTRAN program into object code which performs vector operations.

Simple vector operations such as vector add, multiply, division, etc. and combinations, with no dependences of vector elements on the left- and right-hand side of the equation are easily vectorized. However, even for more complicated operations with an apparent dependency of array elements, the compiler performs a dependency analysis and examines the corresponding loop to see whether array operations can be vectorized safely. Some of the features of today's compilers are, e.g.,

* IF statement conversion,
* In-line expansion,
* Loop distribution,
* Loop fusion,
* Loop interchange,
* Loop unrolling.
* Loop sectioning,
* Reduction function recognition,
* Scalar expansion.

A set of very useful vectorization tools provided by the manufacturers help the user performing a Hot Spot Analysis and optimizing her/his code.

2.2 Parallel Architectures and Parallel Processing

There are basically two main classes of parallel computers: shared memory systems and distributed memory systems. Shared memory systems are widely used today, although mainly as single-processor machines. In parallel mode, a few processors work on one application, all sharing the same main memory. The Fujitsu and NEC machines, the CRAY Y-MP, the new IBM ES/9000, and the mini-supercomputers ALLIANT FX/2800 and CONVEX C240/380 belong to this class.

The parallel computers like the hypercubes Intel iPSC/860, nCUBE/2 and the Connection Machine, the binary tree TX3, Meiko and Suprenum belong to the class of distributed memory systems.

Here every processor has its own local memory (4 to 64 Mbytes) and data is communicated among the processors via a characteristic communication network. These systems are easily scalable in that hundreds or even thousands of identical processors can work on one problem. As a start-up, however, a small system (say 16 to 64 processors) usually is sufficient to implement the code or the application to be solved. The implementation may take several months (or even several years e.g. for very large finite element packages), depending on the structure and size of the code. This has to be validated very carefully in advance; the effort with parallel computers and parallelizations has mostly been underestimated and the resulting performance overestimated, in the past.

Parallel processing on multiprocessor systems is recommended above all for problems which

* need very much CPU (or elapsed) time so that the engineer has to wait a long time for his solution,
* are not treatable at all otherwise.

Parallelization and parallel execution reduces only the elapsed time of a program, but not the overall CPU time (which is the sum of the CPU times spent in all processors), which even increases when using synchronization mechanisms.

Similar to the vectorization of computer programs for the vector computers several years ago, a new dimension in software development has now come up: the parallelization of the programs for the new parallel systems. If such a system consists of several vector processors, vectorization is still necessary. In addition to this, a detailed analysis of the data structure on subroutine or DO-loop level is now absolutely necessary to facilitate a partitioning of the problem into separate parts of nearly equal size which can be processed independently on different processors. Therefore, detailed knowledge of

* the program to be parallelized,
* numerical algorithms in general,
* the special parallel architecture, and of
* the software tools available for parallelization,

is very helpful.

2.3 Shared-Memory Systems

We find mainly three important levels of executing one job on multiple CPUs of a shared memory system in parallel:

* subroutine level,
* loop level,
* statement level.

Parallel processing at subroutine level is called MACROTASKING. Subroutines coded this way are considered tasks being scheduled onto physical CPUs. MICROTASKING is parallel processing mainly at DO-loop level, but there are also several constructs for subroutine and for statement level. With microtasking the user need not explicitly partition the code into tasks but only mark places in the code where multiple CPUs may safely be executing simultaneously. Such code areas are often called "PROCESSES" or "THREADS".

Microtasking results in low programmer overhead and much lower run-time overhead, and is the basis for automatic parallelization or AUTOTASKING.

An important point with parallel processing for machines such as the X-MP, Y-MP, CRAY-2 and 3090-X00 are the tradeoffs that one has to consider when deciding whether to use micro- or macrotasking. First, there is the overhead due to multitasking itself. The biggest OVERHEAD is for creating a task. This is a fact for all multiple CPU machines with a more software oriented (via operating system) parallel processing. Overhead due to microtasking is much lower, in general about one order of magnitude.

The second problem in decreasing run time speed-up of multitasking is the fact that there are serial parts of the program that must use a single processor, and/or must be synchronized for correct communication among different tasks.

2.4 Distributed-Memory Systems

Today, any massively parallel systems are available, such as the Intel i860, nCUBE/2, Connection Machine, Meiko, Suprenum, TX3 and others. It is not possible, here, to discuss all the machines in detail.

In the following, therefore, we briefly discuss the implementation of algorithms on distributed-memory parallel computers. As an example for the discussion, we restrict ourselves to the binary tree computer TX3 of iP-SYSTEMS in Kiel (Germany). Meanwhile, a 64-processor prototype and an 8-processor final version is running and first experiences have been made and results obtained.

The structure of the TX3 is that of a binary tree, each inner node being connected to three neighbouring nodes. The nodes in the lowermost tree level, called leaves, have only one neighbour, while the root has only two. Each node is a father of two sons and/or son of a father. The number of nodes in a tree computer of height t is $2^t - 1$. Because the TX3 is constructed of uniform components, it is easy to increase the number of nodes in powers of 2 and accordingly scale the overall computing power.

The connection of two neighbouring nodes consists of two fast (60 Mbytes/s), 32-bits wide, buffered data channels. The dataflow through each node is controlled by a "communication engine" which works rather independently of the node's CPU, which contains an Intel i860 processor. The data are transferred in a pipelined mode nearly continuously from the source node to the destination. Due to the decoupling of the communication logic from the CPU and due to the buffering of the data channels, a node can communicate and calculate nearly simultaneously.

The TX3, with its tree structure, has a simple and uniform topology. For such structures, the principle of "divide-and-conquer" is one of the possible strategies for parallelization. A problem is successively divided into two sub-tasks of identical complexity. This strategy is realized in programs with the help of the "WAVE" construct. As a rule, a wave proceeds through three phases: In the DOWN part data is divided while flowing from the root to the leaves. In the LEAF part, operations on the data are executed simultaneously in all leaves. Waves are declared and called in a manner similar to that for procedures. The wave concept makes TX3 programs independent of the actual number of processors. For results, see next section.

3. VECTORIZATION AND PARALLELIZATION OF ALGORITHMS

3.1 Vectorization

Production codes in science and engineering applications are usually highly complex. The vectorization of many thousands of statements will naturally frighten every user. With the following strategy, however, an efficient vectorization of a few parts of the program within an acceptable time substantially improves efficiency:

STEP 1: Hot spot analysis with the aid of special software tools (called Flow Trace, Histogram, Spy, Peer, ...) providing a statistical analysis of where computing time is spent in the application code. This information is annotated directly on the compiler listing. Furthermore the DO-loops are marked as vectorized or non-vectorizable.

STEP 2: Vectorization of the most time-consuming subroutines at DO-loop level if autovectorization does not suffice. It should be mentioned here that a highly modular code obtained by structural programming may, in general, be vectorized more easily.

STEP 3: In the case of highly serial algorithms involving, for example, linear or nonlinear recurrences, Step 2 is not successful, and a restructuring or even a complete replacement of the algorithm is necessary.

A "checklist" for efficient vectorization (Step 2) at DO-loop level contains, among others, the following items:

Avoid: IF statements,
Subroutine and function calls,
Memory bank conflicts,
Excessive paging,
Complicated branching within a loop,
Transfer into a loop,
Linear and nonlinear recurrencies,

Use: Many arithmetic operations within vector loops,
Long vectors,
Unformatted I/O,
Chaining.

Some rules for Step 3 are, for example:

Use: Conjugate gradient instead of direct sparse matrix solvers,
Red-black reordering in iteration methods,
Explicit schemes for complicated unstructured grids,
Special library calls to optimized subroutines.

3.2 Parallelization

Efficient serial algorithms are rarely efficient or even optimal for parallel computers. Serial algorithms can be very efficient when they use the latest and global information for the computation of the numerical solutions. However, the first causes data dependencies and recursive data structures and the second one destroys the local character of an algorithm.

Data dependencies and global data access prevent a direct and optimal parallelization. They require synchronization mechanisms which considerably reduce the efficiency if a processor has to wait for data which are updated by other processors. Therefore, it is very important to minimize the number of synchronization points within an algorithm. In this context, so-called asynchronous algorithms with purely local data exchange are superior to algorithms with a more global structure. Problems, for example, related to hyperbolic differential equations, are of a more local character and need therefore only a few synchronizations compared to elliptic problems for which the solution in all the inner grid points depends upon the given solution at the boundary of the computational domain. On the other hand, with time-accurate solutions of engineering and scientific problems, synchronization is necessary after every time step to guarantee the correct solution of the problem.

The best starting point for parallelization is, obviously, the engineering problem itself, which can optimally be cut into pieces and programmed from scratch with regard to the special computer architecture. Here it is highly recommended to carefully organize the program with well-defined modules and correct data transfer mechanisms. A natural subdivision into tasks often results automatically through domain splitting or, for particle problems, through collecting particles to subsets of the same size.

3.3 Example: Restructuring of the SOR-Poisson Solver

Having obtained some useful information from the flow analysis and the vectorization of most of the DO-loops of the central portion of the program accomplished, most of the work seems to be done. Until now, though, we have not yet considered the algebraic structure of the basic algorithm. This is by far the most important consideration. Indeed, a good algorithm poorly coded is usually preferable to a poor one optimally coded. For this reason, we turn our attention to the restructuring of a basic algorithm as a model for more complex finite difference algorithms in computational fluid dynamics, CFD. The results are valid also for many other CFD algorithms.

Consider the Successive Overrelaxation (SOR) method for the following model problem. Let $f(x,y)$ and $g(x,y)$ be continuous functions defined in the interior G and on the boundary dG, respectively, of the unit square $G = \{(x,y)|0 < x < 1, 0 < y < 1\}$.

We are looking for a function $u(x,y)$ continuous in $G + dG$, twice continuously differentiable in G, which satisfies Poisson's equation

$$\Delta u = \frac{\partial^2 u}{\partial x^2} + \frac{\partial^2 u}{\partial y^2} = f(x,y) \quad \text{in} \quad G$$

and the boundary condition $u(x,y) = g(x,y)$ on dG.

We shall primarily be concerned with the linear system of algebraic equations arising from the numerical solution of this model problem using two different five-point difference equations on a mesh of horizontal and vertical lines with mesh spacing $h = 1/(M-1)$ for some integer M. For a given mesh point (x,y) we approximate the second derivatives of the model problem by the usual central difference quotients and obtain, for each interior mesh point, one linear algebraic equation of the form

$$4 * U(I,J) - U(I+1,J) - U(I-1,J) - U(I,J+1) - U(I,J-1) = -h^2 F(I,J).$$

With the ordering I=1,2,...,M of the mesh points in x-, and J=1,2,...,M in y-direction. For the interior mesh points we seek $(M-2)*(M-2)$ approximate solutions $U(I,J)$ of the $(M-2)*(M-2)$ equations forming the linear system

$$Au = b$$

which, for small values of M, can be solved directly with the aid of a special variant of Gaussian Elimination. Usually, however, such systems are solved iteratively, and very often, the Successive Overrelaxation (SOR) method is used in engineering and scientific applications. Therefore, we choose this method as a basis for our discussion of restructuring with respect to special vector computer architectures. As an example, we consider $f(x,y) = 1$ and $g(x,y) = (x^2 + y^2)/4$, with the exact solution $u(x,y) = (x^2 + y^2)/4$. Thus, given the n-th iteration of u, one determines the $(n+1)$th iteration by

$$U^{(n+1)}(I,J) = (1-\Omega)U^{(n)}(I,J) + 0.25\Omega(U^{(n+1)}(I-1,J) \\ + U^{(n+1)}(I,J-1) + U^{(n)}(I+1,J) + U^{(n)}(I,J+1) - h^2F(I,J))$$

for I, J=2,3,...,$M-1$, with the relaxation factor Ω, $0<\Omega<2$, which results by using the most recently obtained approximate values for $U(I-1,J)$, $U(I,J-1)$. It is exactly this feature of the SOR method which makes it, in this form, unsuitable for vector computers: Vector dependences arise for both indices I and J. The algorithm is recursive: $U(I,J)$ depends on $U(I-1,J)$ and $U(I,J-1)$. E.g. for fixed J, running with I from 2 to $M-1$, computing $U(I,J)$ requires $U(I-1,J)$ which is still in the pipelined functional unit.

This dependence is recognized by the autovectorizing compiler and computation is forced to be performed on the scalar processor. Otherwise, on the vector processor, the algorithm would degenerate to the Jacobi method which has a slower convergence rate.

Thus, the original SOR version is no longer a powerful method on vector computers. Therefore, it is necessary to find special structures of the method which fit the architecture of the vector computer and which, furthermore, do not destroy the good convergence property of the method.

A first idea is to change the order of the mesh points into a zebra-like structure. Suppose, in a first half iteration step, the values of U along the odd columns J=const. have already been computed. Then, in a second half step, for J even only the informations of $U(I,J-1), U(I+1,J)$, and $U(I,J+1)$ are used for $U(I,J)$ along each column mesh line J=const.:

$$U(I,J) = (1-\Omega) * U(I,J) + 0.25\Omega * (U(I,J-1) + U(I+1,J) \\ + U(I,J+1) - h^2F(I,J)).$$

Now, along each row mesh line, the final value of $U(I,J)$ is computed as

$$U(I,J) = U(I,J) + 0.25 * \Omega * U(I-1,J).$$

Most of the operations are performed along columns and therefore, data is collected contiguously out of memory. However, the last evaluation is performed running through the memory with very large strides and causing, eventually, memory bank conflicts.

The same procedure is now applied to the $U(I,J)$ along the even columns $J =$const.

As another possibility of vectorization we use red-black ordering of the mesh points (like on a checkerboard).

Consider a checkerboard with red and black areas each representing one point of our computational mesh. Each red point has four black neighbors and each black point has four red neighbors.

Thus, computation of a "red" $U(I,J)$ involves only the four neighbouring black values

$$U(I-1,J), U(I+1,J), U(I,J-1) \text{ and } U(I,J+1),$$

and vice versa. Red-black SOR, therefore, in a first half step computes all the red values using only old black values, and in a second half step computes all the black values using the red values of the first half step. For programming purpose, the following partitioning is more suitable and easily implemented:

```
      DO    2    J = 2,  M-1,  2
      DO    2    I = 2,  M-1,  2
2        U (I, J) = ...
      DO    3    J = 3,  M-2,  2
      DO    3    I = 3,  M-2,  2
3        U (I, J) = ...
      DO    4    J = 2,  M-1,  2
      DO    4    I = 3,  M-2,  2
4        U (I, J) = ...
      DO    5    J = 3,  M-2,  2
      DO    5    I = 2,  M-1,  2
5        U (I, J) = ...
```

where M is an odd integer (in applications, $M-1$ often is a power of 2). However, red-black SOR has several disadvantages, with respect to vector computers, above all

* vector length is only $(M-1)/2$,
* data are gathered with stride 2.

For most vector computers, short vector length and stride 2 reduces performance drastically. Therefore, it is recommended to avoid short vectors and stride 2, if possible. For this, at the very beginning, the red and the black values have to be stored contiguously into red and black vectors, respectively, by hand. But this restructuring, often, is very cumbersome.

In the case of the SOR method, another variant exists which

* avoids stride problems,
* preserves long vectors,
* is easily implemented,
* converges faster than SOR and red-black SOR,

but holds only for the special poisson and potential equations or other problems with a similar finite difference structure. The model problem can be discretized not only by the classical five-points difference equation but also by the following version:

$$\begin{aligned}4 * U(I,J) - U(I+1,J+1) - U(I-1,J+1) \\ - U(I+1,J-1) - U(I-1,J-1) = HF\end{aligned}$$

with $HF = -2h^2 * F(I,J)$, which may be obtained when rotating the classical difference operator. Because of the structure of the difference star ("X"), we will call this method in combination with SOR the XSOR method.

The implementation of the XSOR variant is easy. Only minor changes of the indices are necessary. Now, for fixed index J, running with I from 2 to $M-1$, $U(I,J)$ no longer depends on the previously computed $U(I-1,J)$. Therefore, all values $U(I,J)$ for fixed J can be computed contiguously on the vector processor, as they are stored in main memory. In the following, we will briefly discuss

* ZEBRA-XSOR, without and with loop-unrolling.

For ZEBRA-XSOR the computational mesh is divided into two sets of mesh points, namely red and black points. For J even, all the points are defined to be red, and for J odd, all the points are defined to be black. Then, the algorithm (in principle similar to red-black SOR) looks like

```
      DO   50   J = 2,  M-1,  2
      DO   50   I = 2,  M-1
50       U (I,J) = ...
      DO   51   J = 3,  M-2,  2
      DO   51   I = 2,  M-1
51       U (I,J) = ...
```

For several vector computers, having vector registers as fast storage devices, it is sometimes useful to keep data as long as possible in these registers avoiding time-consuming load and store operations.

Within one ZEBRA cycle, e.g., the elements $U(I,4)$, I=2,3,...,$M-1$, are needed for the computation of $U(I,3)$ and $U(I,5)$. Gathering the computation of the $U(I,3)$ and the $U(I,5)$ into one loop, avoids one additional load of the vector $U(I,4)$. In the literature this is called "Loop-unrolling with a depth of 2". Computing $U(I,3)$, $U(I,5)$, $U(I,7)$ and $U(I,9)$ for I=2,3,...,$M-1$ in one inner loop leads to loop-unrolling with a depth of four and saves three loads (for $U(I,4), U(I,6)$ and $U(I,8)$). The structure of the ZEBRA-XSOR algorithm unrolled with a depth of 2 looks like

```
      DO    50    J = 2,   M-1,   4
      DO    50    I = 2,   M-1
      U (I,J)   = ...
      U (I,J+2) = ...
50    CONTINUE
      DO    60    J = 3,   M-4,   4
      DO    60    I = 2,   M-1
      U (I,J)   = ...
      U (I,J+2) = ...
60    CONTINUE
      DO    61    I = 2,   M-1
61    U (I,M-2) = ...
```

Loop 50 updates all the red values while loop 60 and loop 61 update all the black values. Loop 61 is necessary in the case where $M-1$ is a power of 2. Other modifications are necessary if $M-1$ is not a power of 2.

3.4 Example: Vectorization of Sparse Matrix Vector Products

In the following, as a real example, we demonstrate a detailed optimization procedure for a sparse matrix-vector product of the matrix A with a vector B as it arises typically in finite element codes. The matrix A is given in the so-called compressed storage mode with all zero elements eliminated. Therefore, IX is the index array gathering the corresponding elements of the vector B before multiplying it with the compressed matrix A.

More than 55% of the overall CPU time of the code are spent in this subroutine. As target machine, we used a VAX 9000 with vector board. The original sparse matrix-times-vector subroutine looks as follows:

```
      SUBROUTINE MATRIX (A,B,C,NZ,NS,NC,FACT)
      DIMENSION A (1,1),B(1),C(1)
      REAL*8 A, B, C, FACT, AB, ABA
      IJ=0
      DO 30 I=1,NS
         DO 29 J=I,NS
               ABA=0.
               IJ=IJ+1
               DO 20 K=1,NZ
                  AB=0.
                  DO 10 L=1,K
                     IX=(L-1)*(NZ-0.5*L)+K
                     AB=AB+A(L,I)*B(IX)
 10               CONTINUE

                  DO 11 L=K+1,NZ
                     IX=IX+1
                     AB=AB+A(L,I)*B(IX)
 11               CONTINUE

                  ABA=ABA+AB*A(K,J)
 20            CONTINUE
               C(IJ)=C(IJ)+FACT*ABA
 29      CONTINUE
         IJ=IJ+NC-NS
 30   CONTINUE
      RETURN
      END
```

The SHOW LOOP option of the VAX vector compiler delivers the following loop summary:

```
Loop Summary:  MATRIX

0001
.       Scalar
.
0004
0005      DO I Serial
0006         DO J Serial
0007            Scalar
0008            DO K Serial
0009              Scalar
0010              Vector L
0011              DO L Section
0012                Vector L
0013                Vector L
0014              ENDDO
0015              Vector L
0016              DO L Chunk
0017                Scalar
0018              ENDDO
0019              Scalar
0020              DO L Section
0021                Vector L
0022              ENDDO
0023              Vector L
0024              DO L Chunk
0025                Scalar
0026              ENDDO
0027              Scalar
0028            ENDDO
0029            Scalar
0030         ENDDO
0031         Scalar
0032      ENDDO
0033      Scalar
```

The optimization procedure consists of the following 8 steps:

1. IX depends only upon K, L, but is recomputed for every I, J
   $\rightarrow IX$ in front of the I, J-loops

```
      DO   120      K = 1,NZ
        DO   110      L = 1,K
110         IX (L,K) = (L-1) * (NZ-0.5*L) + K
        DO 111 L = K+1,NZ
111       IX (L,K) = IX (L-1,K) + 1
120     CONTINUE
```

2. ABA is computed for each pair of I, J and added to $C(IJ)$.

This computation can be separated from DO 30 / DO 29, if the I, J-dependency of ABA is taken into account as $ABA(I, J)$.

```
      IJ = 0
      NCA = NC - NS
      DO       130       I = 1,NS
        DO       129       J = I,NS
          IJ = IJ + 1
          C(IJ) = C(IJ) + FACT * ABA (I,J)
129      CONTINUE
        IJ = IJ + NC
130     CONTINUE
```

3. The remaining code is:

```
      DO       41       I = 1,NS
        DO       41       J = I,NS
41      ABA (I,J) = 0.
      DO       30       I = 1,NS
        DO        29        J = I,NS
          DO           20           K = 1,NZ
            AB = 0.
            DO            10            L = 1,NZ
              BIX = B(IX (L,K))
              AB = AB + A(L,I) * BIX
10           CONTINUE
            ABA(I,J) = ABA(I,J) + AB * A(K,J)
20         CONTINUE
29       CONTINUE
30     CONTINUE
```

4. A code analysis shows NS=24, NZ=6, resulting in very short vector lengths (DO 10: NZ=6). Therefore, interchange of the I, J-loops with the K, L-loops. To enable this, AB has to be eliminated. Substitution of AB into $ABA(I, J)$ leads to:

```
ABA(I,J) = ABA(I,J) + A(K,J) *
[A(1,I)*B(IX(1,K)) + A(2,I)*B(IX(2,K)) + ··· + A(6,I)*B(IX(6,K))]
```

and it follows

```
      DO     20     K = 1,NZ
        DO     10       L = 1,NZ
          ABA(I,J) = ABA(I,J) + A(L,I) * B(IX(L,K)) * A(K,J)
10       CONTINUE
20     CONTINUE
```

5. Now the I, J-loops can be interchanged with the K, L-loops:

```
      DO     20     K = 1,NZ
        DO     10       L = 1,NZ
          BIX = B(IX(L,K))
          DO     30       I = 1,NS
            DO     29       J = I,NS
              ABA(I,J) = ABA(I,J) + A(L,I) * BIX * A(K,J)
29           CONTINUE
30         CONTINUE
10       CONTINUE
20     CONTINUE
```

Also, $B(IX(L, K))$ has been moved in front of the I, J-loops. This saves $NS * (NS - 1)$ address computations for every K, L.

6. For fixed I all $ABA(I, J)$ are computed rowwise. A columnwise computation would improve performance for many vector computers. But an interchange of I and J is not directly possible because of $J = I, NS$, which means the following structure:

```
                         o o o o - - - o      I=1
                           o o o - - - o      I=2
                             o o - - - o      I=3
    ABA=                       o - - - o      I=4
                         . . . . . . . . .
                                     - o      I=NS
```

I=1,NS	is the same	J=1,NS
J=I, NS	as	I=1,J

7. In addition, $BIX * A(K,J)$ is independent of I, and it can be moved in front of the I-loop:

```
      DO 20 K = 1,NZ
        DO 10 L = 1,NZ
          BIX = B(IX(L,K))
          DO 29 J = 1,NS
            AIX = BIX * A(K,J)
            DO 30 I = 1,J
30             ABA(I,J) = ABA(I,J) + AIX * A(L,I)
29          CONTINUE
10        CONTINUE
20      CONTINUE
```

8. Because of J=1,NS, the 30-loop is very short for small J. Therefore, a separation of the small vector computations from the larger ones is useful, with the vector-scalar break-even point at J=10, for example:

```
      DO      29      J = 1,NS
        AIX = BIX * A(K,J)
        IF (J .LE. 10) THEN
          DO    30      ISCAL = 1,J
30           ABA (ISCAL,J) = ABA(ISCAL,J) + AIX * A(L,ISCAL)
        ELSE
          DO    31    I = 1,J
31           ABA(I,J) = ABA(I,J) + AIX * A(L,I)
        ENDIF
29    CONTINUE
```

With this, and $ISCAL$ declared as $INTEGER*2$ (for the VAX vector, or similarly with a compiler directive forcing scalar execution), the short loops are computed in scalar mode, and for $J > 10$ the DO 31-loop is executed in vector mode. Now, the optimized version of the program looks as follows:

```
      SUBROUTINE MATRIX (A,B,C,NZ,NS,NC,FACT)
      DIMENSION A (1,1), B(1), C(1)
      REAL*8 A, B, C, FACT, AB, ABA(NS,NS)
      INTEGER*4 IX(NZ,NZ), NCA
      INTEGER*2 KS, LS, ISCAL

            DO   120    KS = 1,NZ
              DO   110    LS = 1,KS
      110         IX (LS,KS) = (LS-1) * (NZ-0.5*LS) + KS
```

```
            DO   111    LS = KS+1,NZ
  111       IX (LS,KS) = IX (LS-1,KS) + 1
  120   CONTINUE

        DO    41     I = 1,NS
          DO    41     J = I,NS
   41   ABA (I,J) = 0.

        DO    20     K = 1,NZ
          DO    10     L = 1,NZ
            BIX = B(IX(L,K))

            DO    29     J = 1,NS
              AIX = BIX * A(K,J)
              IF (J .LE. 10) THEN
              DO    30     ISCAL = 1,J
   30               ABA (ISCAL,J) = ABA(ISCAL,J) +
     %                   AIX * A(L,ISCAL)
              ELSE
                DO    31       I = 1,J
   31               ABA(I,J) = ABA(I,J) + AIX * A(L,I)
              ENDIF
   29         CONTINUE
   10       CONTINUE
   20     CONTINUE

          IJ = 0
          NCA = NC - NS
          DO    130      I = 1,NS
            DO    129      J = I,NS
              IJ = IJ + 1
              C(IJ) = C(IJ) + FACT * ABA (I,J)
  129       CONTINUE
            IJ = IJ + NCA
  130     CONTINUE

          RETURN
          END
```

The SHOW LOOP option of the VAX vector compiler demonstrates our full success for this optimization procedure. The 31, 41 and 129 loops are executed on the vector processor, while for all the other statements, the scalar mode is applied (as was intended):

```
Loop Summary:  MATRIX

0001
.         Scalar
.
0011
0012      DO I Serial
0013        DO J Section
0014          Vector J
0015        ENDDO
0016      ENDDO
0017
.         Scalar
.
0027
0028      Vector I
0029      DO I Section
0030        Vector I
0031      ENDDO
0032
.         Scalar
.
0037
0038      DO I Serial
0039        DO J Section
0041          Vector J
0042        ENDDO
0043        Scalar
0044      ENDDO
0045      Scalar
```

3.5 Parallelization of SOR for Shared-Memory Systems

In order to illustrate parallelization of iterative methods for the solution of algebraic equations arising from the discretization of partial differential equations, we will consider the SOR method for the model problem with a special red-black ordering of the method. For dense system parallelization.

Consider the checkerboard with red and black points (I, J) presented in the last section. Now parallelization of the red-black loops 2, 3, 4 and 5 can proceed in several ways

* task 1: loops 2 and 3; task 2: loops 4 and 5,

* task 1: loop 2; task 2: loop 3; task 3: loop 4; task 4: loop 5,
* all outer loops in parallel,
* all outer loops in parallel chunks,
* with domain decomposition, dividing the two-dimensional region into as many stripes as processors are available, computing the parts of the four DO-loops within the corresponding task.

In all inner loops, however, the vector length was halved and a stride of 2 was introduced with the red-black ordering. Thus, where possible, this kind of restructuring has to be avoided. Therefore, the ZEBRA-XSOR variant of SOR can also be recommended for parallelization.

For ZEBRA-XSOR the computational mesh is divided into two sets (stripes) of mesh points, namely red and black points. For J even, all the points are defined to be red, and for J odd, all the points are defined to be black. Then, the algorithm (in principle similar to red-black SOR) looks like

```
      DO    50    J = 2, M-1,2            DO    51    J = 3, M-2,2
      DO    50    I = 2, M-1              DO    51    I = 2, M-1
   50 U (I,J) = ...                    51 U (I,J) = ...
```

Again, as with red-black SOR, several ways of parallelization for the ZEBRA-SOR method are possible:

* task 1: loop 50; task 2: loop 51,
* all outer loops in parallel,
* all outer loops in parallel chunks,
* with domain decomposition, dividing the two-dimensional region into as many stripes as processors are available, computing the parts of the two DO-loops within the corresponding task.

This time, in contrast to red-black SOR, the vector length within the inner DO-loops are preserved. Therefore, with XSOR, the vector and the parallel potential of the computer is fully exploited. The results in Table 1 for a 4-processor X-MP/48 with 64 banks of ECL memory show that the completely vectorized version (XSOR) is also quite efficiently parallelizable.

Table 1 **MFLOPS and Speed-up Factor for Two Parallel Versions of XSOR on a CRAY X-MP with Four Processors**

N	Unitasking MFLOPS	Microtasking N-2 Tasks/Iterat. MFLOPS	 factor	Microtasking 4 Tasks/Iterat. MFLOPS	 factor
31	77	135	1.7	165	2.1
63	100	244	2.4	300	3.0
127	111	310	2.8	364	3.3
255	115	364	3.2	400	3.5
511	121	451	3.7	453	3.7

Table 2 **Timings in Seconds and Efficiency in % for XSOR for Vector Lengths N=63 and 127 (ES: Event Synchronization, LP: Loop Parallelization)**

XSOR	Procs	ES-LP	Timings N=63	 N=127	Efficiency N=63	 N=127
IBM 3090	1		0.31	5.21		
	2	ES	0.24	2.92	64.1	89.1
	4	ES	0.25	2.17	30.7	59.9
	2	LP	0.28	3.52	56.2	74.1
	4	LP	0.30	3.01	25.5	43.3
CRAY X-MP	1		0.08	1.11		
	4	LP	0.027	0.34	75.0	82.5
CRAY-2	1		0.08	1.23		
	2	ES	-	0.80	-	77.3
	4	ES	-	1.23	-	25.0
CONVEX C220	1		0.33	5.31		
	2	LP	0.22	3.00	75.0	85.7
ALLIANT FX/80	1		1.74	27.48		
	2	LP	0.89	14.02	96.9	98.0
	4	LP	0.48	7.67	90.6	89.6
	8	LP	0.29	4.80	73.4	71.7

Every outer DO-loop is distributed in the first case onto N-2 processes (with DO GLOBAL), and in the second case onto 4 processes (with DO GLOBAL FOR 4). The maximum speed-up factor reached with microtasking is "only" 3.7 because of the IF-test (performed on one processor) to verify the accuracy reached after every iteration step.

Table 2 shows the efficiency $E = Sp/p$ (with speed-up Sp of p processors over one processor) of the different parallel machines is shown for varying granularity (given in number of arithmetic operations). For macrotasking, the double DO-loop is split into 2 or 4 stripes resulting in a granularity of $N * N/2$ (resp. $N * N/4$) times 7 operations. For loop level parallelization, the granularity is only N times 7 operations. Table 3 shows the task granularity for the different problems in number of floating-point operations.

Table 3 Grain Size per Task in Number of Operations for XSOR for Event and Loop Parallelization

N	Event-Paral.		Loop-Paral.	
	2 procs	4 procs	2 procs	4 procs
63	13891	6945	441	441
127	56451	28225	889	889

For the IBM, loop parallelization makes sense only for coarse-grain inner loop bodies with much more than 1000 operations. On the other hand, event synchronization is quite efficient already for grain sizes of the order of 10000 operations.

Because of the heavy use of the control registers, loop level parallelization on the CRAY X-MP is very efficient even on a 4-processor machine.

For the old CRAY-2 with dynamic memory under Unicos 3.0 and with the compiler CFT77 release 2.0, parallelization for up to 3 processors and N=127 was quite efficient. However, when all 4 processors were involved in parallel processing, a dramatic decrease in efficiency was observed, mainly because of memory contention and concurrent operating system activities taking place on one of the 4 processors, and results were not reproducible. For the new CRAY-2 with static memory under Unicos 4.0 with CFT77 release 3.0, efficiency has been improved much.

The hardware oriented dynamic ASAP of CONVEX is very efficient for all granularities above 500 operations and allows for fast job scheduling even in a multiuser environment.

3.6 Parallelization of SOR for Distributed-Memory Systems

To parallelize SOR for distributed-memory systems, the grid is partitioned into rectangular subdomains of approximately equal size which are mapped on the processors of the TX3.

Each block is surrounded by an overlap area which contains the values from the adjacent subdomains. Before each iteration, the values in the overlap area are exchanged with the neighbouring processors. Each processor can then update the points of its associated subdomain without having to communicate with its neighbours during the calculation. For the parallel SOR method we use an improved scheme that staggers exchange steps and calculations of the grid points at the inner boundaries. In this way each grid point is updated using the most recent values of its neighbouring points. Benchmark results for a similar 2-D grid problem on a 64 processor prototype show communication losses of less than 5 %.

3.7 Example: Numerical Grid Generation

The grid generation code TOMCAT computes a 2-D mesh around a configuration in a 2-D rectangular region. The 200-lines kernel approximates a set of two non-linear elliptic differential equations for the grid-point coordinates x, y by an implicit finite difference method. For parallelization, because of the implicit nature of the algorithm, two synchronization points are necessary. The resulting discrete systems of algebraic equations are then solved by a line SOR method. Here the coefficients of the matrices and the tridiagonal systems themselves can be solved in parallel. Between these two steps, however, there is a synchronization point necessary to wait for all matrix coefficients to be completed. Another synchronization point is needed at the end of the iteration step, testing for convergence, so that there are two synchronization points within every iteration. This two-step structure - set up the algebraic system of equations and solution of the system - can also be found in finite element algorithms. In Table 4, timing results and efficiency are given for various shared-memory parallel computers.

A special block version of TOMCAT is MGP: a flexible mesh generation code suitable for parallel computers. The computational domain under consideration is subdivided into blocks (rectangles, cubes) and the iteration is performed in each block independently. For these two codes, TOMCAT and MGP, we consider the basic idea of scalability. In the past, computers often have been compared with fixed sized problems such as Dongarra's 100*100-LINPACK. This measure, however, doesn't make much sense for parallel computers, where fixing the problem size creates a severe constraint, since for increasing number of processors it means that a problem must run efficiently even when its variables occupy only a small fraction of available memory. In practice, problem size increases with the number of processors, and this has to be simulated with benchmarks on parallel computers with many processors.

To demonstrate this idea, we show results of a few simple experiments. At first, we parallelized TOMCAT via domain decomposition, and implemented this version on the ALLIANT FX/80. The results are given in Table 5. The fixed size problem contained 1089 grid points and the scaled problem $p * 1089$ grid points for p processors.

Table 4 TOMCAT, Time in Seconds, Efficiency in %, for Vector Lengths N=63 and 127 (Mac=Macrotasking, Mic=Microtasking and Aut=Autotasking)

TOMCAT	Procs	Task	Time in Seconds N=63	Time in Seconds N=127	Efficiency in % N=63	Efficiency in % N=127
IBM 3090	1		1.51	5.24	-	-
	2	Mac	0.85	2.76	88.8	94.9
	4	Mac	0.57	1.69	66.2	77.7
CRAY X-MP	1		0.71	2.46	-	-
	2	Mac	0.42	1.39	84.5	88.5
	4	Mac	0.26	0.85	68.3	72.4
CRAY-2	1		0.89	3.19	-	-
(Dynamisch)	2	Mac	0.54	1.73	83.1	92.4
	4	Mac	0.60	1.17	36.9	68.2
CRAY-2	1	Mac	0.52	2.08	-	-
(Statisch)	2	Mac	0.32	1.15	81.3	90.4
	4	Mac	0.24	0.67	54.2	77.6
CONVEX C220	1		1.55	5.66	-	-
	2	Aut	1.16	4.16	66.8	68.0
ALLIANT	1		8.06	31.96	-	-
FX/80	2	Aut	4.21	17.53	95.7	91.2
	4	Aut	2.36	10.16	85.4	78.6
	8	Aut	1.61	7.14	62.6	56.0

Although efficiency for the scaled problem decreases much slower than for the fixed size problem, see Table 5, extrapolation of these results would give very poor efficiency for more than eight processors. The main reason for this is the structure of the algorithm, which contains two synchronization points. A modified version which is much better suited for parallel computers is MGP. Here, only one synchronization point is necessary to communicate segment boundaries between neighbouring subdomains.

Results for MGP are given in Table 6 for a fixed problem size of 625 and a scaled size of $p * 625$ grid points on p processors (which is even less than for TOMCAT).

Table 5 Speed-up Factors for TOMCAT for Fixed and Scaled Problem Size on an ALLIANT FX/80

No. of Procs	Speed-up Factor		Efficiency %	
	Fixed	Scaled	Fixed	Scaled
2	1.70	1.94	85.0	97.0
3	2.23	2.84	74.3	94.7
4	2.69	3.71	67.3	92.8
5	2.95	4.20	59.0	84.0
6	3.18	4.50	53.0	75.0
7	3.43	4.90	49.0	70.0
8	3.63	5.01	45.4	62.6

Table 6 Speed-up Factors for MGP for Fixed and Scaled Problem Size on an ALLIANT FX/80

No. of Procs	Speed-up Factor		Efficiency %	
	Fixed	Scaled	Fixed	Scaled
2	1.7	2.0	85.0	100.0
3	2.2	3.0	73.3	100.0
4	2.5	3.9	62.5	97.5
5	2.8	4.8	56.0	96.0
6	3.0	5.6	50.0	93.3
7	3.2	6.4	45.7	91.4
8	3.4	7.0	42.5	87.5

Tables 5 and 6 can only give a rough estimate of the effect of scalability on overall performance, because several other hardware constraints influence efficiency, such as memory and bus contention, and the cache problem. Therefore, TOMCAT has been rewritten also for the distributed memory parallel computer TX3 of iP-SYSTEMS in Karlsruhe. A prototype of this binary-tree machine is now running with 64 processors. Results are given in Table 7 for $p*3500$ grid points. The results of Table 7 demonstrate the potential with highly parallel distributed memory systems.

Table 7 Speed-up Factors and Efficiency for TOMCAT for Scaled Problem Size on the Binary Tree Machine TX3

No. of Procs	Speed-up Factor	Efficiency %
1	1.00	100.00
4	3.99	99.75
8	7.98	99.75
16	15.91	99.44
32	31.70	99.06

4. CONCLUDING REMARKS

When starting parallelizing a program, the user is confronted with many new aspects discussed in this chapter. Some of them are granularity, synchronization, communication, memory contention and parallel algorithms. Some of our experiences made with parallel processing and parallelization are

* The structure of a parallel program is more complicated than its serial counterpart.
* In order to keep the structure of the parallel program simple, only the most time-consuming parts of large programs should be parallelized.
* At the first optimization step the program should be vectorized, because the user has to pay the CPU time for every processor in use. It is then often possible to insert a completely vectorized structure unchanged into the tasks. However, vectorization and parallelization can hamper each other, e.g. if the vector length is reduced when dividing the problems into tasks. A further reduction of the vector length of short vectors (e.g. N=64 elements into $N1$=32 and $N2$=32 elements)

should be avoided with parallelization, because then the efficiency of the processors decreases rapidly.

* Synchronization points have to be inserted where tasks are not equally distributed (being of different granularity) among the processors or have to communicate during execution. Computations are not continued until all the data necessary for the next program steps is available.
* Between two synchronization points the parallel tasks have to be independent and can, therefore, be performed in any order and at any time.
* The task granularity has to be much larger than the overhead due to the serial portion of the program (Amdahl) and the overhead of the generation and synchronization of tasks. Increasing the number of tasks usually results in a decrease of granularity.
* The granularity of the tasks should be (nearly) equal (load balancing). Otherwise, all other tasks have to wait at the synchronization point for the last one to be completed, and the processors are idle.

A highly parallel efficiency can only be achieved on parallel computers in single-user mode. Otherwise, in a heavily loaded multi-user environment, the turn-around time for the job would be unpredictable and efficiency would dramatically decrease. One strategy for multi-user parallel machines resulting in a reasonable efficiency and in a predictable turn-around time is to have a specified number of processors available during run time of the parallel job. This cluster of processors then acts as a single-user parallel machine, and the other machines can be used for a multi-user environment or divided into several other clusters.

A parallel computer used for parallel computations is not a general purpose machine anymore. How the processors are connected to each other is of vital influence on the implementation and on the performance for a specific applications. One application running very efficiently on one parallel system can be quite inefficient on another system.

As can be seen from all this, vectorization is quite common, and for suitable problems also very efficient. However, parallelization today can be very efficient but is not straightforward at all. Therefore, a program should be parallelized only if it is used very often, for real large production runs, and if the algorithm structure is inherently parallel.

The better approach for parallel processing in the future is to take the chance to write completely new code using modern algorithms highly suitable also for distributed memory systems.

VIII UHSNWT INITIATIVE AT NATIONAL AEROSPACE LABORATORY

Masahiro Fukuda, Toshiyuki Iwamiya and Hajime Miyoshi
National Aerospace Laboratory
7-44-1, Jindaiji-Higashimachi, Chofu, Tokyo, 182 JAPAN

SUMMARY

In order to promote CFD (computational fluid dynamics) research activities and to make use of the fruits for the research and development of future aircraft and space planes, it is necessary to develop a CFD-oriented computer (ultra high-speed numerical wind tunnel, UHSNWT) in a couple of years whose processing speed for CFD programs is more than 100 times higher than that of the Fujitsu VP400, and a computer more than 1000 times faster by the end of 1990's.

This chapter is concerned with requirements to the UHSNWT from the viewpoint of the development cost and operation cost, reliability, affinity with CFD programs, operation environment and so on. At the same time, the UHSNWT initiative at NAL is proposed based on

1. the experience of using vector computers,
2. software simulation of the program processing for several configurations of vector computers, and
3. forecast of a future trend of LSI technology.

The discussion concludes that the UHSNWT with performance of more than 100 times higher than the VP400 is feasible with a parallel computer with distributed memory and crossbar network.

1. BACKGROUND OF NUMERICAL WIND TUNNEL

1.1 Present Situation of CFD

The peak performance of vector computers exceeded 1 GFLOPS in the middle of 1980's. At the beginning of 1987, a Fujitsu VP400 with peak performance of 1.185 GFLOPS together with a Fujitsu VP200 was introduced and was put into operation in the National Aerospace Laboratory (NAL). The system allowed the CFD (computational fluid dynamics) research area to be widely extended to not only the numerical simulation of invicid flows so far but also the simulation of flows governed by the Reynolds averaged Navier-Stokes equations (RANS). The VP400 achieves the actual performance of 300 to 700 MFLOPS for RANS simulations with perfect gas. Fruitful results with the power of the VP400 are summarized as follows [1].

(1) Acquisition of reliable data for the flows near design points around clean aircraft configurations (aircraft with no such components as engines, ladders and ailerons).

(2) Acquisition of reliable data for hypersonic flows ($M > 10$) including real gas effects around clean space plane configurations.

(3) Acquisition of reliable data for aerodynamic elements of propulsion systems.

(4) Provision of an effective tool for the research of hypersonic combustion.

(5) Two dimensional flutter analysis including shock separation.

(6) Validation of the methods developed for solving design problems (inverse problems) of aerodynamical elements of aircraft and propulsion systems.

On the other hand, several problems in performing CFD researches have been pointed out at the same time [1,2]. Among these, serious ones are listed below. They are all caused by the insufficiency of computer performance and main memory capacity.

(Q1) The RANS simulation with clean aircraft or space plane configurations needs about 1 million of grid points on the average. With the mesh, it takes CPU time of about 10 hours and main memory capacity of about 200 Mbytes in case of the calculation of transonic flows with the VP400, while it takes CPU time of about 25 hours and main memory capacity of at least about 300 Mbytes in case of the calculation of hypersonic flows involving real gas effects. Accordingly parametric studies, which are key to the development of aircraft, by using RANS simulation seem almost impossible because of too much time needed and high cost. It should be noted that around 10 minutes is a reasonable amount of time for processing each case to allow CFD to be used for parametric studies in the technological development.

(Q2) 5 millions to 15 millions of grid points are required for the RANS simulation with complete aircraft configurations [3]. Once that simulation capability dealing with such complex configurations as mentioned above within a practical amount of time is available, CFD could become as effective a tool for the technological development as wind tunnels. But the performance of computers available at present is insufficient for the purpose.

(Q3) CFD at present stage cannot yet produce reliable data for flows dominated by turbulence phenomena. For this purpose, progress in turbulence modeling should be made. Numerical simulation to establish turbulence models requires calculations with at least 100 millions to 150 millions of grid points. Such calculations are beyond the existing computers' power.

(Q4) Establishment of turbulence models as well as combustion models are indispensable to the improvement of reliability of data acquired by the simulation of flows in RAM/SCRAM jet engines. In this case the number of necessary grid points is estimated almost the same as in (Q3) and is also impossible to handle now.

1.2 From Ultra High Speed Supercomputer to Ultra High Speed Numerical Wind Tunnel

It is expected that CFD would play an important role in the development of future aircraft and propulsion systems especially in the development of space plane, which is now about to start. To meet the expectation, the problems (Q1)~(Q4) listed in the previous section should be solved in advance.

For this purpose,

(R1) In a couple of years a supercomputer which has processing speed 100 times faster than the VP400 and main memory capacity of 32 Gbytes for data storage should be developed and introduced in NAL.

Furthermore, in order to make the most of CFD fruits obtained with the use of a high speed computer with the performance as mentioned in (R1) in the research and development in the field of aeronautical and space technology.

(R2) By the end of 1990s, a super speed computer which has processing speed more than 10 times faster and main memory capacity of more than 10 times larger in fundamentally the same way as the computer with performance (R1) mentioned above should be realized.

On the other hand, the peak performance of the highest-speed commercial supercomputers which will be available on the market around 1993 is estimated 5 to 7 GFLOPS in case of a single vector processor (UP) and 15 to 25 GFLOPS in case of a tightly coupled multi-vector processor (TCMP). There would remain great gaps with our requirements in performance. Because they are of general use, these machines are designed to yield high performance to some extent to wide range of problems and also to have better user interfaces. In this sense we could name "general purpose supercomputer".

In this chapter, we discuss the feasibility of a high speed computer with the performance mentioned in (R1) restricting the purpose to the application to CFD activities. This kind of computer is suitable for the name of an ultra high speed numerical wind tunnel (UHSNWT).

2. DEMANDS IN THE SYSTEM MANAGER'S EYES

In the preceding section we have presented the target performance. But solutions to realize it from the viewpoint of computer engineering are not unique. There are a great number of technical options, even though some of which may be important, some trivial.

We can say that developing a computer with a target performance is nothing but a selection from many options through continuous discussions between computer engineers and user group. Here we give a brief discussion on a couple of technical items which serve as criteria to obtain a solution. These criteria obviously act as external compelling forces to computer engineers. In fact our fixing the target performance of itself is one of the selections done by viewing the present situation and by predicting future developments of computer engineering in consideration of what CFD is like now and is to be like in the future.

2.1 Costs

From the viewpoint of those who are studying and developing aerodynamic technology such as aircraft, the UHSNWT is one of the aerodynamic experimental facilities comparable to conventional wind tunnels. Therefore data acquisition with the UHSNWT should not cost more than with wind tunnels. High cost keeps users from performing such things as parametric studies, to which computers are more suitable.

V. L. Peterson [4] has once said without detail that the cost is no more than 100 dollars per dataset with NASA's main wind tunnels. A detailed investigation on the costs of wind tunnel experiments in the U.S.A. has been reported by National Research Council [5].

Since one of the main objectives of the UHSNWT is to make use of RANS simulations in the research and development of aerodynamic technology for aircraft and so on, we should investigate the way not only how to realize the performance but how much such costs can be reduced as the development, production and operation costs. Then the costs per dataset with wind tunnels should be taken into consideration. The time has come when data costs as well as data reliability of RANS simulations should be considered.

Although it is outside the scope of this report to discuss the development, production and operation costs, we here give a brief sketch about LSI technology on which the hardware development expenses and operation costs of the UHSNWT greatly depend.

The experience in the past shows that mass-production effects have drastically reduced the costs of memory chips and logic elements (DRAM is a good example), while the costs of chips which is not to be mass-produced are slow to decrease (typically in high-speed ECL-RAM). Generally speaking, memory chips line up in order of cost per bit in the following (this is also the order in terms of speed and power consumption).

$$\text{ECL-RAM} > \text{BiCMOS-SRAM} > \text{CMOS-SRAM} > \text{DRAM}$$

The cost of 256 Kbits CMOS-SRAM with access time of 35 nsec has come down significantly due to mass-production effects, while the cost of 64 Kbits ECL-RAM with access time of 15 nsec has been reluctant to lower. Therefore it is most advantageous in terms of costs that products of general use on hand are employed for the UHSNWT. Even if such products are not fitting and a new LSI has to be developed, it is needed to reduce the cost by spreading it to other fields and stimulating its mass-production.

Since power consumption of LSI affects profoundly the operation cost, it may be allowable to adopt as a criterion in choosing LSI whether expected power consumption per average effective processing speed falls within admissible range. Figure 1 shows the transition of power consumption per gate of major logic elements used in domestic supercomputers. Judging from the trend the power consumption per gate of ECL would not be far below 1mW at the end of 1992. It seems reasonable that it would fall between 1.2 and 1.3 mW. This is one of the factors to judge whether ECL chips are to be used in the UHSNWT.

Furthermore it may happen that high-speed memory chips consume 10 times as large power per bit as medium-speed ones. We also need to choose low power chips for main memory not only in operation but on standby .

2.2 Reliability

Although the recent advances of LSI and packaging technology have exercised a great influence on the progress of reliability of computers, it is well known that measures taken for the improvement of reliability may often go against the increase of performance and cost reduction. It is possible to improve gradually the reliability of software after the system installation, while the reliability of hardware is determined essentially in the planning stage.

It is 3 years and 11 months since the VP400 was introduced in NAL. It has been in continuous operation all day long except maintenance time of 3 hours per month and 17 holidays in total around the New Year's day and in summer. This implies that the system runs for 8310 to 8320 hours per year with regular maintenance time of 30 to 40 hours (once a month). User programs spend CPU time of about 7900 hours per year which forms about 95 % in the total operation time.

Under these circumstances, no hardware fault has occurred for about 4 years but 2 double bit errors in the main memory which took place just after the installation. In case of the Fujitsu VP series the number of gates used for guaranteeing the reliability and maintainability such as circuits for error-check and diagnosis is estimated 20 % to 30 % in the total number of gates.

How much quantity should be spared for the reliability and maintainability, even though the UHSNWT would need much more hardware than the VP400. The UH-

SNWT is not a mere experimental computer but a computer to use for the research and development of CFD technology. UHSNWT users might take strict attitudes toward errors or system downs since they are not experts of computer engineering but CFD researchers and engineers in general. To maintain high-level reliability is one of the keys to success as well as to realize the target performance. Therefore it is necessary to give a careful consideration how much should be invested for hardware reliability improvements adjusting the balance with the performance and cost.

If one bothers to calculate doubly due to the doubt about the data obtained or if one often has to repeat the same calculation because of errors, the high target performance does not make any sense. Although there is no doubt that too much investment for reliability should be avoided, it is a matter of engineers moral whether they attack sincerely the maintenance of reliability in the course of development.

3. THE UHSNWT INITIATIVE

3.1 Starting Point

From the experience we have in the utilization of vector computers, we begin our examination of the UHSNWT with the following suppositions.

(P1) Vector computers are suitable for processing CFD programs except CFD involving rarefied gases.

(P2) Vector computers with vector registers (VR) are good systems. FORTRAN compiler technology for this type of computers is highly sophisticated.

(P3) Such standard hardware functions as list vector processing and vector compression/expansion processing of vector computers coupled with high level FORTRAN compiler help CFD programs to achieve the vectorizing ratio of about 99%. The vectorizing ratio increases according to the scale of problems.

The VP series has a timer to measure how long each user job makes use of the vector unit. According to the statistics obtained so far there are few CFD programs whose average vectorizing ratio is less than 95 %. Almost all CFD programs are assessed to have vectorizing ratio of more than 99 %.

Briefly speaking, large scale CFD programs imply a large number of grid points from the viewpoint of computers. Based on this fact it seems reasonable to adopt a following additional supposition at the starting point.

(P4) The parallelization ratio of about 99% can be achieved for CFD programs under the vectorizing ratio of about 99% if a suitable parallelism is assumed.

3.2 Hierarchical Structure of the UHSNWT Memory

The biggest issue in the development of high speed computers in recent years is how to supply data into pipelines whose speed is getting higher and higher. Higher performance of a vector computer is attained by the combination of speed-up of LSI chips, increase of the number of vector elements processed concurrently with a pipeline (we call multiplicity of a pipeline), increase of the number of pipelines executable in parallel (these pipelines are running independently), and increase of the number of processors. But it is not easy to have a high speed main memory of large capacity commensurate with the theoretical peak performance.

Strategies to solve this problem are twofold.

(i) Arrange the main memory consisting of many banks in the use of high-cost and high-power consumption memory chips in order to assure high data transfer rate.

(ii) Arrange the main memory consisting of many banks in the use of medium-cost and medium-power consumption memory chips.
In this case, the lack of data transfer rate are supplemented by VR of large capacity.

From the user's viewpoint, these solutions have their own merits and demerits, and different conclusions may be derived depending on his attitude to high speed computers.

Judging from the future trend of LSI technology, the gap between processing speeds and data transfer rates of main memory is likely to widen so that the problems may not be solved if we rely only on the strategies mentioned above. In particular, it is the case if we attempt to make processing capability more than 100 times and then 10 times further.

To clarify the situation we are in, we make a rough estimation.

The fundamental speed of a vector computer is determined by the pipeline pitch time (τ_p). Figure 2 shows that the transition of τ_p of vector computers. The horizontal axis indicates the year of the first delivery.

From Fig. 2, τ_p in 1992 to 1993 is expected to be at most 2 nsec. Now we configure the UHSNWT pipelines by suitably combining 160 add and 160 multiply pipelines with τ_p=2 nsec together with multi-processor architecture. In this case the UHSNWT has a peak performance of 160 GFLOPS (divide pipelines are ignored since divide operations occurs rarely).

Under this assumption, how much transfer rate is needed for main memory? The use of such VRs of large capacity as in the VP400 may reduce the burden of main memory to supply data. If the ratio of data transfer capability to peak performance is assumed to be the same as that of the VP400, data transfer rate of 640 Gbytes per second is required. If it is the same as that of the VP200, data transfer rate of 1.28 Tbytes per second is required. If it is the same as that of the CRAY-YMP, data transfer rate of 2.56 Tbytes per second is required. Although it is surely impossible to have a main memory of tens of Gbytes with data transfer rate of 2.56 Tbytes per second in 1992 or 1993, a main memory with data transfer rate of 640 Gbytes per second might be possible if the cost and power consumption are ignored. But it might be impossible to meet requirement (R2) on data transfer capability, more than 10 times higher, in the same manner by the end of 1990's.

Therefore we try to make up for the shortage of data transfer capability by changing the main memory hierarchical configuration as in Fig. 3. A parallel computer as in Fig. 3 is called distributed memory parallel processor (DMPP) or multi-computer (MC).

This type of computer has main memory distributed to each processor element (PE) as the name indicates. If the data needed for calculation with each PE are stored on the corresponding distributed memory, the insufficiency of transfer rate of main memory as mentioned above can be solved more easily.

The main memory hierarchy consists of 3 stages, VR, main memory distributed on one PE (primary memory) and main memory distributed on the other PEs (secondary memory).

Let us denote the peak performance of the UHSNWT by PP and the number of PEs by N (the performance of each PE is PP/N). Then the data transfer rate needed between VR and the primary memory is reduced to the one corresponding to the peak performance PP/N, while the requirement for transfer rate capability between the primary memory and secondary memory can be much reduced if the data required for the calculation are so allocated that the access to the secondary memory might occur rarely.

In the case of DMPP, the main memory capacity distributed to each PE (denoted by PEM) is the sum

$$PEM = M/N + R_1 + R_2 + R_3,$$

where M is the total amount of data needed to store on main memory, R_1 is the amount for operating system kernel, resident at each PE, R_2 is the amount for CFD program procedure and R_3 is the amount for data storage received from other PEs. R_1 and R_2 might be at most several Mbytes.

Now we will make a brief comment on R_3. In the case of DMPP, in processing CFD programs total computational domain (denoted by CD) is divided into subdomains and each PE is attached to a subdomain, which is denoted by CD_i when the PE is PE_i. The domain on which the data required for processing the subdomain CD_i lie is denoted by DD_i. Then

$$CD = \bigcup_{i=1}^{N} CD_i$$

and in general

$$CD_i \subset DD_i.$$

Set

$$AD_i = DD_i - CD_i,$$

and assume that

$$AD_i \subset CD_{j_1} \cup CD_{j_2} \cup \cdots \cup CD_{j_m}.$$

for some $j_1, j_2, \cdots, j_m$. Then PE_i must communicate with PE_js ($j = j_1, j_2, \cdots, j_m$) to receive the data required. Hence it is necessary to keep the amount and frequency of data transfer between PEs as little as possible in order to make efficient use of DMPP. R_3 is the amount of data which belong to and are derived from AD_i. R_3 depends not only on schemes, algorithms and programming technology CFD programs use but also on compiler technology taking into account the performance of interconnection network. Therefore R_3 could be large. We evaluate that R_3 is in between one fourth and half the amount M/N of main memory.

From now on, we assume that the UHSNWT is of DMPP type as illustrated in Fig. 3.

3.3 Required Performance – From the Manager's Viewpoints

In the previous section we have determined that the UHSNWT is of DMPP type. In this section we discuss the required performance of each PE.

In order to make ordinary users, not computer experts, familiar with a parallel computer,

(i) the establishment of a standard of FORTRAN including expressions on parallelism and a compiler with capability of parallelization and optimization for programs described with the FORTRAN standard, or a compiler with capability of parallelization and optimization for programs described with the conventional FORTRAN, and

(ii) the development of debugging and tuning software for supporting parallel programming,

are necessary.

In view of the years taken for the refinement of programming environments for vector computers which seems much easier to achieve, the establishment of good programming environments for parallel computers is a hard and time-consuming task. One way to solve the problem is to furnish computer experts with an experimental parallel computer to promote the establishment. Another way, which we take, is to introduce the UHSNWT among CFD researchers and engineers and then to promote not only the improvement of programming environments but also the spread of knowledge useful for high level usage of the parallel computer through discussion between computer specialists and ordinary users.

When the latter way is taken, at least the following must be guaranteed to UHSNWT users.

(i) A job which is suitable to be processed with the VP400 can be executed with 1 PE and has full compatibility at source program level.

(ii) 1 PE should deliver higher performance than the VP400 in executing the job as in (i).

(iii) The UHSNWT should be a heterogeneous parallel computer, which has system control PEs performing system management and external communication.

(iv) The system control PEs should have the same scalar units as PEs for calculation (vector computers with VR) to prevent two different PEs from being developed. Therefore the performance of the scalar units should be higher than that of the VP400. The performance in this degree is at least necessary for the control of the UHSNWT. By virtue of the system control PEs, the UHSNWT can be directly connected to high-speed LAN without any special host computer.

(v) Cooperation of specialists should be guaranteed for the parallelization of programs for technical development of aerodynamics.

(vi) Cooperation of specialists should be guaranteed for the parallel programming for large-scale jobs.

It is clearly favorable in terms of cost and power consumption if 1 chip processor can be employed as PE. It is predicted that 1 chip processor will attain the equivalent performance with the VP400 in about 2000 [6]. Hence we cannot adopt 1 chip processor as PE. We expect that 1 chip processor would have been developed for the UHSNWT with the performance mentioned in (R2).

Since each PE should have higher performance than the VP400, the number of PEs comprising the UHSNWT is estimated to be at most about 200. Then a powerful interconnection network can be constructed, and it is possible to make the UHSNWT suitable for the various CFD methods which will be mentioned below.

The UHSNWT does not always execute one job by using all PEs exclusively. The UHSNWT often performs multi-job operation by assigning each job to some of PEs, (this does not mean the multi-job with one PE) and hence the UHSNWT is necessarily a so-called MIMD parallel computer.

3.4 Configuration of PE

3.4.1 Speed-up of PE

As mentioned before, vector computers at present employ for speed-up

(i) increase of the number of vector elements processed concurrently in a pipeline, (Fujitsu VP series, NEC SX-2, 3 series).

(ii) increase of the number of pipelines executable in parallel, (HITACHI S810/820 series, NEC SX-3 series).

(iii) increase of the number of PEs of tightly coupled multi-processor type, (CRAY XMP/YMP series, NEC SX-3/44).

If (iii) is adopted, each PE is constructed with several vector computers and a hierarchical structure of processors is introduced. Although there is no reason to avoid the introduction of such hierarchical structure, to do this, it is necessary that the multi-vector computer of TCMP type should be fully transparent to FORTRAN users and that the compiler technology should be available which allows to achieve the performance to the highest extent. We realize that the compiler technology for the multi-vector computer of TCMP type does not reach this level. Under these circumstances, we think it is too early to introduce the hierarchical structure of the processors.

(ii) implies, for example, the introduction of several, say 2, multiply pipelines into one PE. In this case the performance will be improved for some programs, but as a whole it is not effective for the amount of hardware invested. In order to improve the effectiveness, if the number of pipelines is as few as 2, loop unrolling can be employed. However the capacity of the VR, data transfer rates of the VR and between the main memory and the VR should be doubled. Considering the pros and cons, we do not adopt this method for the UHSNWT.

3.4.1.1 Pipelined vector computers with large VR

(i) In 1977 NAL introduced the FACOM 230-75 AP, which had been jointly developed with Fujitsu since 1972. This vector computer has 1792-word programmable local high-speed memory which is accessible from load/store and arithmetic pipelines. It has been verified that the burden on the data transfer rate of the main memory is reduced by making effective use of this memory.

(ii) In 1978, NAL started the numerical simulator project which enabled us to process 3D wing RANS simulation in about 1 hour. The Fujitsu VP400 has been developed with the VP200 as its basis, based on a NAL's assertion that large scale VR not only reduces the burden on the main memory data transfer capacity but also allows the multi-element pipelines to be made effective use of. The VP400 demonstrates performance in the range of 300 to 700 MFLOPS in the CFD program processing as supports the NAL's assertion. The performance is as good as those of other high-end supercomputers of the same generation.

The main features involving the VP400 performance are the following.

machine clock time (τ) = 13.5nsec,
pipeline pitch time (τ_P) = 6.75nsec,
start-up time for add/multiply pipeline = 12τ,
start-up time for divide pipeline = 36τ,
multiplicity of add/multiply pipeline = 4,
load/store pipeline 1, data transfer rate 8×8 bytes/τ,
start-up time for load/store pipeline (access time),
31τ for load
18τ for store
bank cycle time of the main memory,
7τ for load
20τ for store.

It is said that the VP400 shows low effective performance compared with the peak performance. It is because the pipeline start-up time is too long compared with the multiplicity of pipelines, data transfer rate is relatively small compared with the peak performance and there is only one load/store pipeline, as is seen in the above features. This kind of computers display the defect when processing DO loops consisting only of vector operations such as

$$a = b + c \times d : \qquad a,\ b,\ c \text{ and } d \text{ are vectors.}$$

But in CFD programs the ratio of the processing time for such DO loops is small, while the processing time for the DO loops involving a large amount of vector operations occupies much larger part in the total processing time. In processing DO loops involving successive vector operations, high-level compiler and proper vector instruction controls allow the pipelines to execute consecutive vector processing in an overlapped fashion with large-scale VR and appropriate VR access controls provided that there is no conflict. Hence the actual start-up time can be much smaller than the value listed in the catalogue.

We will explain this by taking a multiply pipeline as an example. In Fig. 4 VPS_i, VOI_i and VPE_i respectively stand for

VPS_i: the time when the pipeline begin the action, that is the time when the first pair of vector elements is put into the multiply pipe.

VOI_i: the time when the result for the first pair is output.

VPE_i: the time when the vector multiplication ends, that is the time when the result for the final pair of vectors is output.

The start-up time of the multiply pipeline is defined by

$$VST_i = VOI_i - VPS_i.$$

(We have defined this way for simplicity. The start-up time for add/multiply pipeline of the VP400 is 10τ in this sense.) The theoretical start-up time of the multiply pipeline is denoted by TST and the actual start-up time by EST_i. Figure 4-1 shows the case where vector multiply 1 and 2 are performed independently, while Figure 4-2 shows the case where 2 vector multiplies are performed in an overlapped fashion. In Fig. 4-1,

$$EST_1 = VST_1 = TST, EST_2 = VST_2 = TST.$$

In Fig. 4-2, the start-up time for vector multiply 2 is

$$VST_2 = VOI_2 - VPS_2 = TST.$$

But if vector multiply 1 and vector multiply 2 take place as in Fig. 4-2,

$$EST_2 = VOI_2 - VPE_1 < TST.$$

If 2 vector multiplies are completely consecutively executed, $EST_2 = 0$. In any case we have for all i

$$0 \leq EST_i \leq TST.$$

The average effective start-up time $AEST$ for the multiply pipeline in DO loop processing is given by

$$AEST = \frac{\sum_{i=1}^{n} EST_i}{n},$$

where n is the number of vector multiplications appearing in the DO loop. The same is true for add , divide and load/store pipelines.

Actual $AEST$s estimated by using the VTAP, a vector processing time chart analyzing program for vector computers, will be shown in the next subsection.

3.4.1.2 VTAP simulation

(1) VTAP

The VTAP is a program to analyze the operation of a vector computer by producing the timing diagram of DO loop processing from object programs generated by compilers. Functions of the VTAP for the VP series are as follows

(i) It can simulate precisely instruction controls from vector instruction fetches to the executions.

(ii) It can simulate accurately VR access controls.

(iii) It can simulate precisely the operations of all arithmetic pipelines.

(iv) It simulates main memory access by using simple main memory conflict models for load and store, consecutive vector access, access with a stride and indirect access respectively.

(v) It simulates scalar instruction processing by simplifying. Since most of the scalar instructions in DO loops are address calculations, the simplification does not affect the simulation accuracy.

In order to analyze other computers than the VP series, the VTAP has additional functions that allow to treat various configurations of pipelines and to change such parameters as performance, start-up times, the number of pipelines executable in parallel and modeling parameters. It also can deal to some extent with other methods of VR access control and instruction execution control.

(2) Selection of DO loops for VTAP simulation

Selection of typical workload for the performance estimation of computers is one of the difficult works at present. The criterion for selecting workload depends on standpoints and viewpoints. Based on the fact that processing objects of the UHSNWT are limited to CFD and most of the users are expected to be engineers and researchers in the community of aerospace industries and universities in Japan, we first have investigated CFD programs and chosen about 1000 DO loops from them. We then have selected 18 of them by considering

(i) the number of arithmetic vector instructions,

(ii) the ratio of the number of add/subtract instructions to that of multiply instructions,

(iii) the ratio of the number of arithmetic vector instructions to that of load/store instructions,

(iv) whether load/store operations contain accesses with a stride or indirect accesses,

(v) the processing speed of the VP400.

We have not selected the DO loops only because their processing time occupies a large part in the total processing time. For our objective is to clarify the performance of computers in processing DO loops of various kind. And we also have not selected DO loops typically used in each algorithm of CFD programs because compilers at present especially for vector computers with large VR resolve each equation into a series of elementary calculations and optimize the instruction order and hence the order of equations does not make any sense.

3.4.2 VTAP simulation results

Table 1 shows the VTAP simulation results. In the table, α stands for the performance estimated by the VTAP (VTAP MFLOPS), β stands for the measured performance by executing the DO loops with the VP400 specifying vector length of 128 (measured MFLOPS) and hence α/β means the VTAP simulation accuracy. L, S, A, M and D in vector instructions column denote the number of load, store, add/subtract, multiply and divide operations respectively contained in the DO loop. The effective start-up time column shows the respective average start-up times ($AEST$) given in clock time of load/store, add, multiply and divide pipelines in executing the DO loop.

Table 2 shows the VTAP simulation results with vector length of 32. We observe from Table 1 and 2 the followings.

Table 1 Estimation of Processing Speed with Vector Length of 128

				vector instruction					effective startup time			
No	α	β	α/β	L	S	A	M	D	L/S	A	M	D
1	256	254	99.2	107	80	31	82	2	2.92	11.61	6.68	18.00
2	815	700	85.9	21	50	176	170	4	6.06	2.91	1.69	15.00
3	961	870	90.5	110	59	300	340	5	12.49	1.41	1.38	36.00
4	686	654	95.4	28	39	140	169	8	9.94	3.85	2.36	24.50
5	777	720	92.6	70	15	113	117	6	7.35	4.41	2.82	11.33
6	467	414	88.7	26	18	143	147	1	14.14	5.34	3.56	36.00
7	388	402	103.6	6	7	4	17	1	7.15	12.00	1.41	36.00
8	785	870	110.8	22	4	35	33	0	9.88	2.29	2.67	—
9	305	341	111.8	7	4	6	9	1	7.65	11.33	8.00	36.00
0	309	354	114.7	22	2	14	20	2	4.29	10.29	7.60	36.00
11	433	420	97.1	42	2	36	54	2	5.30	9.11	5.11	22.00
12	259	256	98.7	30	10	20	10	0	3.72	9.00	12.00	—
13	458	498	108.7	25	5	20	15	0	3.47	6.00	9.60	—
14	583	605	103.7	31	5	25	27	0	3.22	5.92	5.04	—
15	254	237	93.3	13	5	9	4	0	4.44	9.33	12.00	—
16	408	417	103.5	32	6	24	42	0	3.58	10.00	5.14	—
17	596	560	93.9	34	45	202	202	7	11.72	4.41	3.32	20.00
18	701	705	100.5	67	8	200	162	7	9.03	3.20	3.58	21.14

(i) In the case with vector length of 128, *AEST* of the load/store pipeline is much smaller than the theoretical start-up time. In the case with vector length of 32, *AEST* is comparatively large, but is also smaller than the theoretical start-up time and is about 10 or less except in the cases DO loop 3, 6 and 11. The weighted average of the theoretical start-up times with respect to the numbers of load and store instructions in DO loop 3, 6 and 11 turns out to be 26.5, 25.7 and 30.4 respectively. By comparing these values with the *AEST* listed in Table 2, we see that the *AEST*s in these cases are also much smaller than the theoretical start-up times.

On such vector computers as VP400, with the aid of by optimization technology of compilers, vector elements required for the execution are consecutively loaded in advance on the VR and most of the temporary data generated through vector operations remain on the VR. The occurrences of load or store operations between the VR and the main memory is restricted to the case where VR capacity is insufficient and where the results obtained in the DO loop need store operations. This is the reason why we obtain such results as in Table 1 and 2.

Table 2 Estimation of Processing Speed with Vector Length of 32

No	α	effective startup time			
		L/S	A	M	D
1	114	5.12	12.00	8.29	18.00
2	325	5.94	6.17	5.09	16.00
3	314	15.72	9.81	8.69	36.00
4	278	10.76	7.85	6.23	26.00
5	321	9.76	7.42	6.30	23.33
6	185	18.17	8.53	7.74	36.00
7	155	6.85	12.00	4.71	36.00
8	312	11.35	6.29	6.91	—
9	111	11.64	12.00	10.22	36.00
0	103	12.88	12.00	10.00	36.00
11	159	14.68	11.33	7.78	36.00
12	95	6.65	10.40	12.00	—
13	164	7.63	9.00	9.07	—
14	212	7.89	9.60	9.78	—
15	97	6.72	11.11	12.00	—
16	180	9.55	10.33	7.33	—
17	236	11.81	7.47	7.28	18.29
18	297	12.48	7.04	6.70	16.57

(ii) Most of *AEST*s of add and multiply pipelines are smaller than the theoretical start-up time even though the ratio are not as high as in the case of load/store pipelines. Especially *AEST*s in many cases with vector length of 128 are much less than the theoretical start-up times. When the number of add/subtract and multiply operations in the DO loop lack balance, *AEST* of one pipeline with more operations always tends to be small compared with the theoretical start-up time, while another *AEST* tends to approach the corresponding theoretical value. This tendency implies that in most of the cases, vector data required for the vector operation already exist on the VR when the vector instruction is issued.

(iii) Most of the *AEST*s of divide pipeline are the same as the theoretical start-up time. This is because the frequency of vector divide is comparatively so rare that the divide pipeline acts as in Fig. 5-1.

From the observations above we claim that the following assertion on multi-element pipelined vector computerswith large VR be established.

In processing DO loops containing large amount of vector operations, which take the most part in CFD programs, large VR can prevent the decline of effectiveness of multi-element pipeline and the drop of processing capability due to only one load/store pipeline.

3.4.3 PE model and its feasibility

In this section we first choose some models as candidates of the UHSNWT PE. The processing speed of each PE model for CFD programs is estimated with the VTAP simulation results for the 18 DO loops discussed in the previous section. Judging from this estimation together with the tendency of the recent development of logic VLSI and their costs, we decide the configuration and performance of PE and what type of logic VLSI to use.

3.4.4 PE models

We begin with limiting PE models to be considered.

(i) Assume that the start-up time of add and multiply pipelines be 10 τ, and that the start-up time of divide pipeline be 26 τ. Two of the arithmetic pipelines, add/subtract, multiply and divide, are executable in parallel. The load/store pipelines are always executable in parallel.

(ii) The multiplicity of each pipeline (PM) is assumed to be 1, 2, 4, 8 or 16. If PM is m, m results are obtained every τ in case of add or multiply pipelines, while $m/7$ results in case of divide pipelines. This is the same feature of divide pipeline of the VP400. In the case of load/store pipelines, m of 8bytes data can be transferred every τ.

(iii) The number of load/store pipelines is one of the following 3 patterns, $L/S \times 1$, $L/S \times 2$, and $L \times 2 + S \times 1$. Here $L/S \times 2$ means 2 pipelines which can be used for both load and store operations. The other symbols have similar meanings. Then in case that $PM = 8$ and $L \times 2 + S \times 1$, 16 of 8 bytes data load and 8 of 8 bytes data store can be done every τ.

(iv) The number of VRs is 16 or 128.

(v) The start-up time of load/store pipeline is listed in Table 3.

Table 3 Start-up Time of Load and Store Pipeline

access type	A	B
consecutive vector load and vector load with a stride	20τ	30τ
indirect vector load	25τ	35τ
consecutive vector store and vector store with a stride	10τ	15τ
indirect vector store	15τ	20τ

PE model is constructed as one of the combinations of (i) to (v). In the simulation, we assume that the access control of vector register and the control system from instruction fetch to execution are the same as those of the VP400. We use the same scalar operation and main memory access conflict model as those used in the VTAP simulation of the VP400. Since the accuracy of the VTAP simulation has been clarified for the VP400 in the previous discussion, the VTAP can simulate to some extent such PE models as set above. Object codes generated by compilers for the VP series are employed as input data of the VTAP.

Then a question arises whether the FORTRAN compilers for the VP series can generate optimized code for different pipeline configuration and different number of VRs. We can say yes to the question since

(i) The FORTRAN compilers for the VP series have reached at considerable level.

(ii) As for the number of VRs, the FORTRAN compilers for the VP series have a function of assigning the number of VRs by inserting a directive before the DO loop and can generate optimized code according to the assignment.

(iii) We have used the FORTRAN compiler for the VP400 in the case of $L/S \times 1$ and the FORTRAN compiler for the VP200 in the case of $L/S \times 2$. Accordingly we have obtained optimized codes in both cases.

There seem to remain some problems in the case of $L \times 2 + S \times 1$. But for DO loops with a great number of vector operations, load and store operations occur separately with each other in time and are made overlapped with vector arithmetic operations in the optimized code. Hence we can substitute the object code for $L/S \times 2$ case.

It should be noted that the performance may be slightly underestimated for DO loops including a small number of operations.

Here there seem to remain some problems in adopting the same control method from instruction fetch to execution and the same main memory access conflict model as of the VP400. Because if the control system and main memory performance for the PE are improved, the simulation for the VP400 might underestimate the performance of the PE model. But since the underestimation implies the estimation at the safer side for the simulation purpose, this is not so important.

Table 4-1 Estimation of PE Models

vector length= 128												
	S1	S2	S3	S4	S5	S6	S7	S8	S9	S10	S11	
PM	1	2	4	8	16	8	8	8	8	8		
$\tau(ns)$	4	4	4	10	20	10	10	10	10	10		
*	a	a	a	a	a	b	c	a	b	c	α	β
**	B	B	B	B	B	A	A	A	B	B		
VR	128	128	128	128	128	16	16	16	128	128		
No. 1	144	275	506	349	271	263	330	169	549	555	256	254
No. 2	369	733	1449	1112	800	640	673	532	1116	1117	815	700
No. 3	432	860	1702	1321	855	528	594	423	1338	1338	961	870
No. 4	326	646	1263	951	670	540	557	442	959	962	686	654
No. 5	373	734	1426	1067	725	677	712	568	1155	1156	777	720
No. 6	242	475	915	654	461	554	579	468	668	670	467	414
No. 7	237	449	815	550	371	564	562	564	550	548	388	402
No. 8	419	802	1520	1099	761	868	921	687	1189	1186	785	870
No. 9	198	373	670	445	305	524	521	462	502	500	305	341
No.10	203	387	704	472	322	563	575	481	538	555	309	354
No.11	229	448	857	604	442	747	746	607	725	724	433	420
No.12	175	329	587	386	277	590	639	387	565	613	259	256
No.13	276	523	949	640	479	905	916	645	868	878	458	498
No.14	343	659	1192	812	594	1048	1072	817	1040	1050	583	605
No.15	168	314	556	362	257	549	576	366	514	544	254	237
No.16	212	413	787	566	431	831	836	563	776	777	408	417
No.17	296	582	1129	827	595	568	599	480	846	848	596	560
No.18	340	674	1313	970	708	543	571	460	999	1001	701	705

*:L/S pipes a:$L/S \times 1$ b:$L/S \times 2$ c:$L \times 2 + S \times 1$ **:startup time
alpha:VTAP MFLOPS *beta*:measured MFLOPS

Table 4-2 Estimation of PE Models

vector length= 32							
	S12	S13	S14	S15	S16	S17	S18
PM	1	2	4	4	8	16	
$\tau(ns)$	4	4	4	10	10	20	
*	a	a	a	a	a	a	α
**	B	B	B	B	B	B	
VR	128	128	128	128	128	128	
No. 1	125	215	337	135	170	102	114
No. 2	316	586	982	393	456	248	325
No. 3	425	826	1068	427	443	224	314
No. 4	284	532	871	348	403	215	278
No. 5	324	610	1035	414	468	246	321
No. 6	214	388	568	227	267	144	185
No. 7	180	314	479	189	232	128	155
No. 8	357	638	1015	406	450	231	312
No. 9	168	278	381	152	173	91	111
No.10	176	295	402	161	180	94	103
No.11	214	378	552	221	248	127	159
No.12	147	241	347	139	169	90	95
No.13	237	400	598	239	255	134	164
No.14	298	507	743	297	325	168	212
No.15	139	226	321	128	153	83	97
No.16	197	357	571	228	278	146	180
No.17	257	471	724	290	338	185	236
No.18	311	584	938	375	431	226	297

*:L/S pipes a:$L/S \times 1$ **:startup time
alpha:VTAP MFLOPS

3.4.5 Analysis of VTAP simulation

Table 4-1 and 4-2 show the results of VTAP simulation for several kinds of PE models. The case with vector length of 128 is shown in Table 4-1, while the case with vector length of 32 in Table 4-2. The VTAP simulation has been performed for 18 cases from S1 to S18. In the tables No. 1 to No. 18 indicate the number of the DO loop used. The same DO loop are employed in Table 4-1 and 4-2. The main features of the DO loops are listed in Table 1. The machine clock period τ in the table is temporal for the calculation of MFLOPS. If τ is reduced to $1/R$ times, then MFLOPS increases to R times.

The reason why only DO loops which appear frequently in CFD programs are not used in the VTAP simulation for determining a PE model is as follows. The PE of the UHSNWT has specification of greater processing speed in processing CFD programs than the VP400. To meet surely the specification, it is necessary that PE is faster than the VP400 for DO loops of various kind contained in CFD programs at NAL.

In the following we first confirm that the selected DO loops are not biased for estimating the effect of speed-up of PE. Next we select a PE model by using the VTAP simulation results.

(1) Firstly we confirm that the selection is not biased with respect to the effects of pipeline multiplicity on the processing speed of each DO loop.

Let i be the number of a DO loop and l be the vector length. Set

$$D(i)EPMm_1/m_2(l) = S_1/S_2; \quad i = 1, 2, \cdots, 18.$$

S_1; the processing speed (MFLOPS) of DO loop i with $PM = m_1$

S_2; the processing speed (MFLOPS) of DO loop i with $PM = m_2$

Without loss of generality, we may assume $m_1 > m_2$.

What distribution of $D(i)EPMm_1/m_2(l)$ can lead to the conclusion that the selection is not biased? The upper bound of $D(i)EPMm_1/m_2(128)$ might be m_1/m_2. The upper bound of $D(i)\ EPMm_1/m_2(32)$ is surely lower than that of $D(i)E\ PMm_1/m_2(128)$. We may consider approximately as the lower bound, $EPMm_1/m_2(l)$, the ratio of the pipeline processing speed with $PM = m_1$ to that with $PM = m_2$ assuming that the pipeline processing is performed independently as illustrated in Fig. 5-1. We have 5 kinds of pipelines, load, store, add, multiply and divide. Among these, divide pipeline can be excluded because the frequency of vector divide operations is relatively low. The VTAP simulation discussed in 3.4.2 shows that $AEST$ of the load/store pipeline of the VP400 is much smaller than the theoretical value and is almost the same as the theoretical start-up time of add pipelines. Assume the same thing holds for the UHSNWT PE models with large VRs. Then $EPMm_1/m_2(l)$ for add pipelines may be taken as the lower bound of $D(i)EPMm_1/m_2(l)$.

Since the theoretical start-up time for add pipelines is assumed 10 τ, the vector add processing time T_v with $PM = m$ and vector length of l is

$$T_v = (10 + l/m)\tau.$$

Hence the calculation of T_v for the case with $PM = 4$ and $PM = 1$ yields

$$EPM4/1(128) = 3.29,$$

$$EPM4/1(32) = 2.33.$$

S1 and S3 in Table 4-1 show the DO loop processing speeds in MFLOPS obtained under the same condition except pipeline multiplicity. Hence dividing MFLOPS in S1 by MFLOPS in S3 yields $D(i)EPM4/1(128)$. In a similar way $D(i)EPM4/1(32)$ is obtained from MFLOPS in S12 and S14. These values are shown in Table 5. We see that $D(i)EPM4/1(128)$ is distributed uniformly in the range between 3.31 and 3.94, while $D(i)EPM4/1(32)$ is distributed uniformly between 2.27 and 3.19. The upper bound of $D(i)EPM4/1(32)$ is about 20 % less than that of $D(i)EPM4/1(128)$, while the width of interval on which $D(i)EPM4/1(32)$s lie is about 50 % wider than that of $D(i)EPM4/1(128)$. From the distribution of $D(i)EPM4/1(l)$, we can conclude that the selection is not biased with respect to the effects of pipeline multiplicity.

Table 5 Effects of the Multiplicity of a Pipeline

	$D(i)EPM4/1$ (128)	$D(i)EPM4/1$ (32)
No. 1	3.51	2.70
No. 2	3.93	3.11
No. 3	3.94	2.51
No. 4	3.87	3.07
No. 5	3.82	3.19
No. 6	3.78	2.65
No. 7	3.44	2.66
No. 8	3.63	2.84
No. 9	3.38	2.27
No.10	3.47	2.28
No.11	3.74	2.58
No.12	3.35	2.36
No.13	3.44	2.52
No.14	3.48	2.49
No.15	3.31	2.31
No.16	3.71	2.90
No.17	3.81	2.82
No.18	3.86	3.02

(2) Next we confirm that the selection is not biased with respect to the effects of performance improvement of main memory, that is the performance improvement of load/store pipelines in PE models.

Table 6 Effects of the Improvement of Main Memory

	EM_1	EM_2
No. 1	1.59	1.95
No. 2	1.00	1.27
No. 3	1.01	1.30
No. 4	1.01	1.26
No. 5	1.08	1.25
No. 6	1.02	1.24
No. 7	0.996	0.996
No. 8	1.08	1.34
No. 9	1.12	1.13
No.10	1.18	1.20
No.11	1.20	1.23
No.12	1.59	1.65
No.13	1.37	1.42
No.14	1.29	1.31
No.15	1.50	1.57
No.16	1.37	1.48
No.17	1.03	1.25
No.18	1.03	1.24

When the performance of main memory is doubled or tripled with the performance of pipelines fixed, DO loops may be processed more quickly but the possibility of the processing speeds being doubled or tripled is likely to be small for CFD programs. This holds especially for PE models with large VRs. We investigate the effects of improvement of main memory performance with S4 and S10 when the number of VRs is large and with S7 and S8 when the number of VRs is small. The results are displayed in Table 6. In the table, EM_1 is the ratio of the DO loop processing time (MFLOPS) in S10 to that in S4, while EM_2 is the ratio of the DO loop processing time (MFLOPS) in S7 to that in S8. In the case of large VRs, we see that our DO loops contain those for which the improvement of main memory takes little effect (less than 15 %), quite effect (between 15 % and 50 %) and much effect (more than 50 %). When the number of VRs is as small as 16, the improvement of main memory is expected to have quite effect on the speed-up of DO loop processing. In fact the processing speeds are improved by 20 % to

50 % for most of the cases as expected. But there are 2 cases where the improvement of main memory takes little effect and 2 cases where 60 to 100 % improvements are acknowledged. The features of the DO loops selected are of all kinds. Hence we can conclude that the selection is not biased with respect to the effect of the main memory improvement.

(3) From the discussion above on the selection of DO loops we may adopt the following as a criterion for estimating the processing speed of the UHSNWT PE. PE must process all the 18 DO loops selected faster than the VP400.

In order for the processing speed of PE to be higher than that of the VP400 for all the DO loops with $L/S \times 1$ and case B in Table 4, the machine clock period (τ) should be less than

$$\tau = 1.6\text{nsec} \quad \text{for } PM = 1$$

$$\tau = 3.1\text{nsec} \quad \text{for } PM = 2$$

$$\tau = 6.3\text{nsec} \quad \text{for } PM = 4$$

$$\tau = 12.6\text{nsec} \quad \text{for } PM = 8$$

$$\tau = 12.9\text{nsec} \quad \text{for } PM = 16$$

with safety ratio of 10 %.

It follows from Table 4 that in the case of 16 VRs and PM less than 8, more processing speed than the VP400 for all the DO loops can not be expected even if the performance of load/store pipeline or main memory is improved. On the other hand, in the case of 128 VRs, the gain by virtue of the improvement of main memory or load/store pipeline appears to be small as a whole. The improvement of main memory is one of the most effective methods to make computers achieve the performance to some extent even when treating vectors of short length, DO loops with a small number of operations and programs of low vectorizing ratio. But since the UHSNWT does not deal with such programs, we do not adopt this type of PE models in view of the cost for improving the main memory.

3.4.6 LSI chips for PE

Figure 5 and Figure 6 show the tendency of speed and density of the recent logic chips. From these we can expect that

(i) ECL gate array with 30000 gates/chip, power consumption of 40 w/chip and gate delay time of 50 psec,

(ii) BiCMOS gate array with 100000 gates/chip, power consumption of 20w/chip and gate delay time of 350 psec, and

(iii) CMOS gate array with 200000 gates/chip, power consumption of 10 w/chip and gate delay time of 600 psec,

would be available in a couple of years. Power consumption per gate is 1.3 mW for ECL gate array, 0.2 mW for BiCMOS gate array and 0.05 mW for CMOS gate array.

Although there is no room for doubt that CMOS logic chips are most advantageous in terms of both power consumption and density, the machine clock time is limited to around 20 nsec unless we think of such cooling system as the liquefied nitrogen cooling. But as discussed in the previous section, clock time of 20 nsec does not meet the performance requirement for PE even if $PE = 16$. Furthermore since the machine clock time of scalar units of the VP400 is 13.5 nsec, the scalar performance is clearly inferior to that of the VP400. This does not meet the performance requirement for the system control. For the reasons above we cannot adopt CMOS gate array.

If BiCMOS gate array in (ii) is used, the machine clock time could be made between 8 nsec and 12 nsec, presumably PE with clock time of 10 nsec might be realized. Accordingly PM can be 8 when BiCMOS gate array is employed in the PE.

If ECL gate array in (i) is used, it is difficult to make PE with machine clock time of 1.6 nsec. Hence PE with $PM = 2$ and τ=3 nsec is reasonable and may be achieved.

Which should be chosen, ECL or BiCMOS? We estimate the number of gates needed by considering the total number of gates per load/store and arithmetic/logical pipelines with $PM = 1$ as unity. Then the total number of other gates including 128 Kbytes VR and 64 Kbytes cache memory (gates for main memory and interconnection network are excluded) is about 10. This is a rough estimate by assuming that RAM & Logic chips be employed for memory such as VR and scaling to the number of gates with logic chips of equivalent power consumption. Then the total number of gates comes to 12 with ECL gate array because $PM = 2$, and 18 with BiCMOS gate array because $PM = 8$. Doubling the number of gates with BiCMOS gate array for comparison on the same basis as that with ECL gate array, the total number of gates for BiCMOS gate array amounts to 36.

On account of power consumptions per gate of each gate array as mentioned above the ratio of the power consumption with BiCMOS gate array to that with ECL gate array turns out to be 16 to 7.2. Hence it is more advantageous if BiCMOS gate array is employed in PE. Although this is a rough estimate on power consumption, both chips should be considered equal in terms of power consumption if significant difference does not appear in such rough estimate.

Next we compare the two from the viewpoint of cost. ECL gate array being considered here is the one employed in high-end large-scale computers and commercial supercomputers. On the other hand, BiCMOS gate array is the one used in low-end large-scale computers and medium or small scale computers. This means that BiCMOS gate array is produced massively. Therefore the cost per gate of BiCMOS gate array may be lower than that of ECL gate array.

In terms of reliability, BiCMOS gate array might be more favorable since chip count needed is less.

Although the cost in fact depends on the activity of market and is ambiguous, we think that the above discussion might be true on the whole. Hence we decide to use BiCMOS gate array in PE. If we take τ=10 nsec, the scalar performance of the UHSNWT is higher than that of the VP400 since the machine clock time of the VP400 is 13.5 nsec.

3.5 Configuration of Main Memory

3.5.1 Realization of target main memory capacity

Figure 7 shows the recent development of memory chips. The progress of memory chips is being made favorably as is expected. 1 Mbits SRAM and 4 Mbits DRAM are now entering mass-production stage and 4 Mbits SRAM and 16 Mbits DRAM will be also put into mass production line within a couple of years. On the other hand, recently delivered supercomputers FACOM VP2000 series and NEC SX-3 series have maximum main memory capacity of 2 Gbytes, which are made with 20000 1 Mbits SRAM and

80000 256 Kbits SRAM respectively. From the discussion in 3.2, we know that main memory capacity (PEM) of PE of DMPP type is estimated by the formula

$$PEM = M/N + R_1 + R_2 + R_3$$

where M is memory capacity for storage of CFD data and N is the number of PEs and that R_1 and R_2 are at most several Mbytes and R_3 is $(0.25 \sim 0.5) \times (M/N)$. Assuming that both R_1 and R_2 are 5 Mbytes in total and using M defined by the specification (R1) of the UHSNWT, the total main memory capacity TM of the UHSNWT amounts to

$$\{(0.25 \sim 0.5) \times 32 + 0.005 \times N\} \text{ Gbytes.}$$

If we take $N = 200$, TM lies in the range between 40 Gbytes and 50 Gbytes, which is $20 \sim 25$ times as large as those of the supercomputers mentioned above. If we employ 1 Mbits SRAM for the main memory of the UHSNWT, the chip count amounts to between 400 thousands and 500 thousands. If we employ 4 Mbits DRAM, the chip count is in between 100 thousands and 120 thousands. Taking the speed into account, we should employ SRAM. Compact packaging of 400 thousands to 500 thousands of chips might be possible considering the present state of the art and future advances expected of packaging technology for supercomputers. Accordingly the feasibility of the UHSNWT main memory in a couple of years can be considered high.

In terms of costs, 1 Mbits SRAM will rapidly become more advantageous by means of mass production effects. Hence we decide to adopt 1 Mbits SRAM as main memory chips of the UHSNWT.

Next we have to decide the capacity and configuration of main memory of PE. If the number of PEs is set 200, the main memory capacity of each PE needs 200 Mbytes to 250 Mbytes. In view of the requirement (i) in 3.3 we fix the main memory capacity to be 256 Mbytes.

To evaluate the performance of main memory, we assume that SRAM with access time of 20 nsec be employed. Then the reasonable bank cycle time τ_m is about 4τ. Data transfer rate of the load and store pipelines is 8×8 bytes/τ. If the data transfer rate of interconnection network is assumed 4 bytes/$\tau \times 2 \sim 8$ bytes/$\tau \times 2$, main memory with 64 way interleave might be enough. It has been recently reported in [11] that with the use of a trick on memory chip technology, the bank cycle time can be reduced half compared with usual memory chips. If we apply a similar trick to memory chips, the bank cycle time would decrease to about 2τ, and hence the number of interleave ways could be reduced as small as 32.

Mass production of such memory chips is considerably profitable for the main memory configuration and performance of the UHSNWT.

3.5.2 Affinity with CFD programs

The transition of recent CFD algorithm can be seen as follows from the viewpoint of parallel computers.

(1) Schemes used for calculation tend to be more complex for pursuing higher accuracy.

(2) Researches on unstructured mesh as well as structured mesh are making progress. Although easy applicability to complex configurations is an advantage of unstructured mesh, there seem unsolved problems, for example, how to apply turbulence models. Structured mesh dominates the CFD community in Japan for the present. This is the case especially for the application to practical problems in the research and development of aerodynamics. The authors cannot predict the future trend in this line.

(3) Implicit methods, especially the IAF method among these, are more dominant than explicit methods. If researcher's concern tends towards such problems as unsteady problem for which implicit methods cannot be much favorable, the situation at present might change. But for the time being we cannot ignore the fruits obtained with the IAF method in the field of practical applications.

(1) demands much from the data transfer capability of interconnection network of parallel computers. List vector processing is required in processing CFD programs using unstructured mesh as mentioned in (2). This demands much from not only the data transfer capability of interconnection network but hardware and software functions. Solving linear equations with an implicit method requires sequential methods (Gauss elimination for example) in the direction of one axis in 3D space. Although attempts to modify sequential methods for parallel computers are being made in many places, reliability on the convergence, accuracy and range of applicability seems insufficient.

If one applies sequential sweep methods in the direction of one axis, say x, on structured mesh it is reasonable that one direction, say y, is used for vectorization, and the remaining direction, say z, for parallelization. In this case the number of PEs is enough with at most 1000 in view of the scale of problems the UHSNWT is supposed to handle. About 200, which is the number of PEs we have decided, is favorable in this sense.

Since the direction of sequential sweep changes from x, y to z direction in turn in the IAF method, not only the data transfer capability but also the topology of interconnection network become important factors. The UHSNWT must deal effectively with all of the CFD methods mentioned at the beginning of this section. Hence it is not desirable to adopt an interconnection network with which the performance fluctuates between high and low as CFD methods are changed. Various kinds of network topologies have been proposed so far and the selection of interconnection network is a key factor in the research and development of parallel computers.

Among these, it is fully connected network or crossbar network that may meet our requirements. They are most flexible and have rare conflicts during data transfers. Fully connected network has no conflict on the network. But connecting 200 PEs this way needs several millions of wires. The connection may be impossible unless the transfer bit width is extremely reduced. One disadvantage of a fully connected network with small bit width is that most paths are idle when data traffic concentrates on some paths between specific PEs (for example between neighboring PEs) while data transfer rate between them is insufficient because of the small bits width.

No conflict occurs on crossbar network if data transfer is one-to-one. It also has an advantage that the distance between every pair of PEs is all the same as in fully connected network. It had been believed that the maximum number of PEs which can be connected with a crossbar network was at most in the range between 16 and 32. Recently it has been reported in [12] that 128 PEs can be connected with a crossbar network. There 8 bits width is considered. The parallel use of 8 crossbar networks of this type leads to 64 bits width. Hence it might be possible to adopt a crossbar network in the UHSNWT.

Data transfer rate of a crossbar network for the UHSNWT needs at least about 4 bytes/τ × 2. This is one eighth of the data transfer rate between VR and main memory, which is 8 bytes × 8/τ. If one can double the rate to 8 bytes/τ × 2, there may be no worry on the data transfer rate of network. It is also satisfactory that the access time of crossbar network is not so large compared with other networks.

4. OVERALL HARDWARE CONFIGURATION OF THE UHSNWT

4.1 Summary

The block diagram of the UHSNWT is illustrated in Fig. 8.

PE_i; processing element ($i = 1, \cdots, 200$)

CP_i; control processor ($i = 1, \cdots, m$, $m = 2 \sim 4$)

Performance of PE;

clock time τ=10 nsec
peak performance: 1.6 GFLOPS
main memory capacity: 256 Mbytes
data transfer rate between main memory and VR: 6.4 Gbytes/s

Overall performance of the UHSNWT;

clock time τ=10 nsec
peak performance: 320 GFLOPS
total main memory capacity: 51.2 Gbytes
data transfer rate between main memory and VR: 1.28 Tbytes/s

Network performance;

data transfer rate: 0.8 Gbytes/sec/PE $\sim$ 1.6 Gbytes/sec/PE
total data transfer rate: 160 Gbytes/sec $\sim$ 320 Gbytes/sec

External memory;

capacity: 8 Gbytes
data transfer rate: 1.6 Gbytes/sec $\sim$ 3.2 Gbytes/sec

We have evaluated the performance of PEs for the 18 DO loops with the VTAP. Here we will estimate with the VTAP how much performance is shown for general CFD programs. CFD programs employed here are followings.

NS3D: 3 dimensional RANS simulation program using a TVD scheme solved with the

IAF method. This program features that each DO loop is so modified as to include large vector operations matching the VP400. The vectorizing ratio is 99.8 %.

NSMD: Three dimensional RANS simulation program using a multi-zonal method and a TVD scheme solved with the IAF method. Since this program is developed to employ for the aerodynamic analysis of subsonic, transonic and supersonic flows around the unmanned space plane named HOPE the multi-zonal method is used to treat the complex configuration, the HOPE connected with the H-II rockets. It is a program for general vector computers, that is it is not tuned up especially for the VP400. The vectorizing ratio is 99.4 %.

Table 7 Estimation of Processing Speeds with the VTAP in MFLOPS

program	UHSNWT PE VTAP	VP400 VTAP	VP400	VP200 VTAP	VP200
NS3D	996	704	649	—	—
NSMD	751	511	483	289	285

Processing speeds for the CFD programs are evaluated through the following procedures.

(1) Denote the total processing time measured with the VP400 by T_A, the total processing time of all DO loops measured with the VP400 by T_D and the total processing time of all DO loops estimated with the VTAP by T_{VD}.

(2) The total processing time T_{VA} is evaluated by

$$T_{VA} = T_{VD} + (T_A - T_D).$$

(3) Calculate MFLOPS by dividing the total number of floating point operations in all DO loops by the total processing time T_{VA}. Hence in estimating the processing speed with the VTAP, we do not take into account that the PE of the UHSNWT is faster than the VP400 for scalar operations.

4.2 Reliability of the UHSNWT

As mentioned in 2.2, reliability is a very important factor. Suppose that the same amount of hardware is invested for reliability as in the VP400 and that MTBF of each PE is X hours. It follows that one of the PEs may fault every X/N hours when the number of PEs is N. Hence it is necessary to prepare for errors by creating a backup copy of the intermediate results into high-speed solid external memory every time when the number of PEs employed times the elapsed time amounts to X hours. We estimate the capacity of the external memory needed for this purpose. Suppose that the number of variables which require a backup copy such as aerodynamic variables and coordinates is 8 and the number of mesh points is 150 millions. Then at least

$$150 \text{ Mbytes } \times (5 \times 4 \text{ bytes } + 3 \times 8 \text{ bytes }) = 6.6 \text{ Gbytes}$$

is required. Since the external memory has other purposes such as pre-staging and de-staging for the system, at least 8 Gbytes may be necessary.

4.3 Overall Performance

In estimating the performance of the PE of the UHSNWT, we have assumed the same instruction execution control system and main memory model as of the VP400. There are some points to be improved.

(i) access control of VR.

(ii) the number of pipelines which can operate in parallel. In the VP400, 2 of the add, multiply and divide pipelines can operate concurrently. This limitation sometimes causes bottleneck.

(iii) start-up times of pipelines.

If the improvements are made, the performance of the UHSNWT would be made higher than that presented in this report since the bank cycle time of main memory is already improved considerably.

Now in view of Table 7 we assume that the performance of the PE of the UHSNWT is 1.4 times higher than that of the VP400 and the efficiency of the PE is 60 % taking into account the high performance of the crossbar network. We conclude that the overall performance of the UHSNWT with 200 PEs could be

$$1.4 \times 200 \times 0.6 = 168$$

times as high as that of the VP400.

4.4 Feasibility of UHSNWT Meeting Requirement (R2)

We begin with a forecast of the technology required.

(i) In view of Fig. 5 and Fig. 6, at the end of 1990s CMOS gate array would contain some millions of gates. BiCMOS gate array would contain at least one million of gates.

(ii) The cost of 16 Mbits SRAM would be sufficiently reduced. 64 Mbits SRAM would be at the first stage of mass production.

(iii) With the advancement of compiler technology one would be able to use vector computers of TCMP type as easily as vector computers of UP type.

(iv) Crossbar network would be able to connect 512 PEs.

Supposing (i) through (iv) we take as an processing node 2 PEs coupled tightly with each other with machine clock time of 4 nsec, pipelines with multiplicity of 8 and main memory capacity in between 1 Gbytes and 2 Gbytes. Connecting 512 nodes would result in peak performance of 4 TFLOPS and main memory capacity of 500 Gbytes through 1 Tbytes. Then the UHSNWT would have processing capability more than 10 times as much as the UHSNWT of first generation in dealing with CFD programs. Compared with the UHSNWT of first generation the memory chip count of the PE would become $1/4 \sim 1/2$ if 16 Mbits SRAM is used as a memory chip and the logic chip count would become 1/5 if BiCMOS is used as a logic chip. Hence the total size would be halved.

The UHSNWT of second generation meeting requirement (R2) would be realized with high expectancy.

5. CONCLUDING REMARKS

G. H. Hegel declared in Gruntlinien der Philosophoe des Rechts on occurrences of historical events.

"Was vernunftig ist, das ist wirklich; und was wirklich ist, das ist vernunftig"

If one believes in the absolute spirit in the sense of Hegel, the complementary proposition above would equally hold. But there seems few who believe in the absolute spirit in the world. In the computer world, we often encounter something whose raison d'etre are doubtful at first sight but it turns out to have a certain rationality in a worldly sense with the investigation in detail. The second proposition of Hegel is true in a limited sense. How about the first one?

We have been pursuing the rationality from many viewpoints-budget, management, CFD, computer technology, NAL's research activity and so on during the technical investigation of the UHSNWT and have presented the feasibility of a UHSNWT.

Would the rationality be realized? Since the world does not change according to the logic of the absolute spirit, the UHSNWT could not be realized if we assume indifferent attitude. Quite a long time has elapsed since the research and development of parallel computers began. There is no parallel computer for CFD in the world whose performance is much higher than that of "general purpose supercomputer". Our investigations predict that its realization is coming close at hand. Once one thing is realized many followers is sure to appear. Believing that the UHSNWT would give a powerful driving force over the aerospace activities in the world, we have to struggle towards the realization.

Authors would like to thank Mr. M. Takamura at Fujitsu, Mr. M. Furukatsu at NEC and Mr. S. Kawabe at Hitachi for helping with information on the tendency of technologies and for fruitful discussions. Authors also would like to thank Mr. N. Nishi at NEC for his paper on pipelined memory.

6. REFERENCES

[1] Hirose, N. and Isogai, K., "Numerical Aerodynamics Simulation Technology for Aerospace Engineering (in Japanese)," The Japan Soc. for Aeronautical Sciences, Vol 38, No 443, Oct., 1990, pp. 507-515.

[2] Miyoshi, H., "Computer Requirements for the CFD Development (in Japanese)," NAL SP 13, Sept., 1990, pp. 1-26.

[3] Tanioka, T., "Status and Outlook of CFD Technology at Mitsubishi Heavy Industries Nagoya (in Japanese)," NAL SP 13, Sept., 1990, pp. 69-81.

[4] Peterson, V. L., NASA CP 2032, Feb., 1987, pp. 5-30.

[5] NASA "Influence of Computational Fluid Dynamics on Experimental Aerospace Facilities A Fifteen Year Projection," NRC Washington D. C., March, 1983.

[6] Gelsinger, P. P., " Microprocessors circa 2000," IEEE Spectrum Oct., 1989,

[7] Miyoshi, H. et al., "Vector Processing Time Analysis Program," NAL TR to appear in 1991.

[8] Fujitsu, "Private Communication," Oct., 1990.

[9] Fujitsu, "Private Communication," Oct., 1990.

[10] Hitachi LTD, " Private Communication," Oct., 1990.

[11] Nishi, N., et al., "Pipelined Storage for Vector Processors," 4th International Conference on Supercomputing and 3rd World Supercomputer Exhibition, May, 1989.

[12] Murakami, K. et al., "The Kyushu University Reconfigurable Parallel Processor – Design of Memory and Intercommunication Architectures," Proceeding of ACM SIGARCH 1989 International Conference on Supercomputing (ICS 89), 1989.

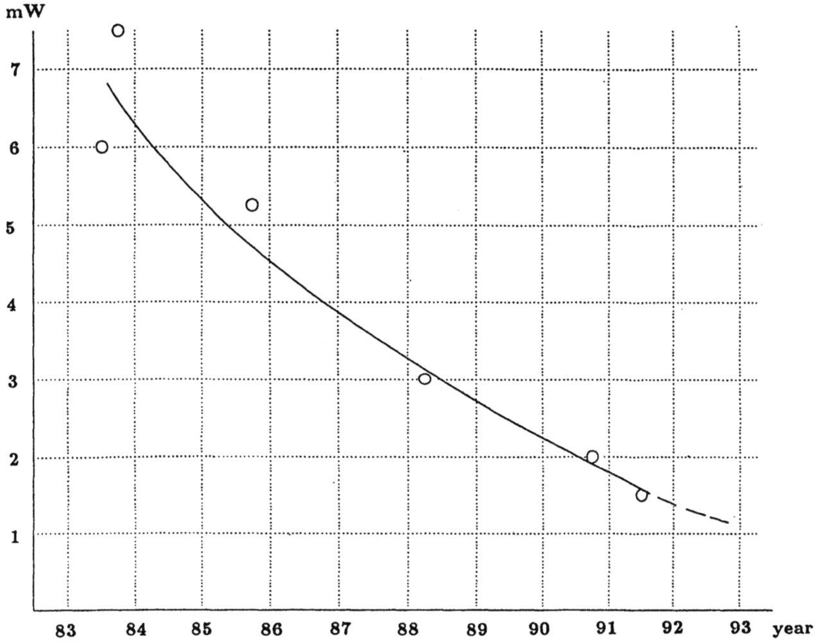

Fig. 1 Transition of the Power Consumption per Gate of Logic

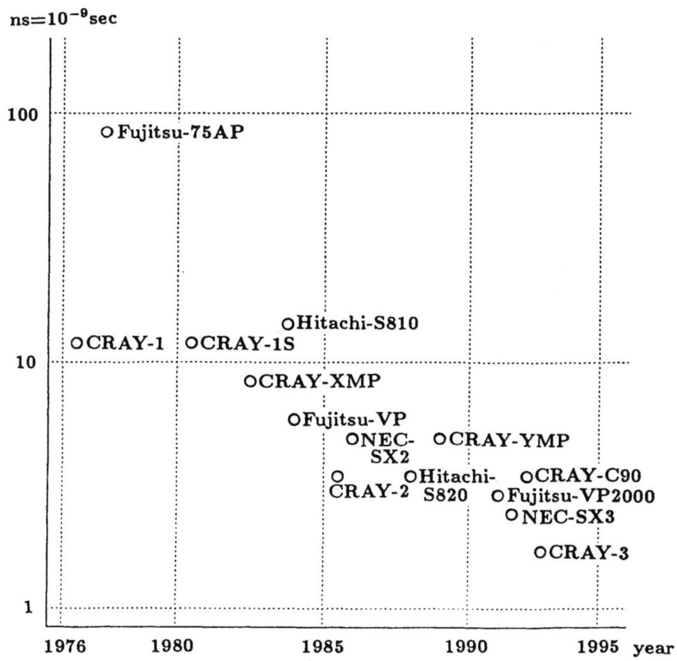

Fig. 2 Transition of the Pipeline Pitch Time

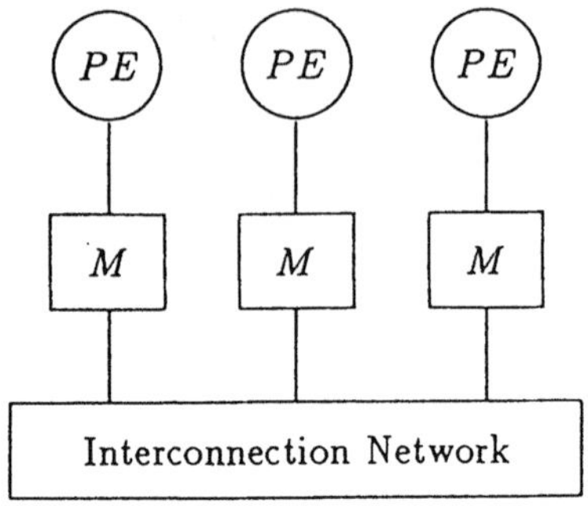

PE; Processor Element
M ; Main Memory

Fig. 3 Distributed Memory Parallel Processor

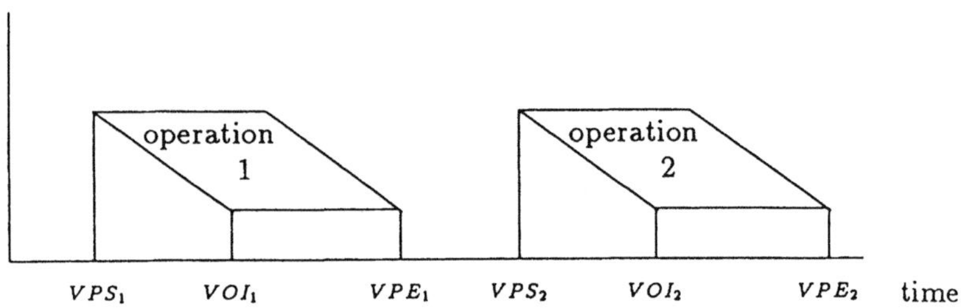

Fig.4-1 Start-up Time for Independent Operations

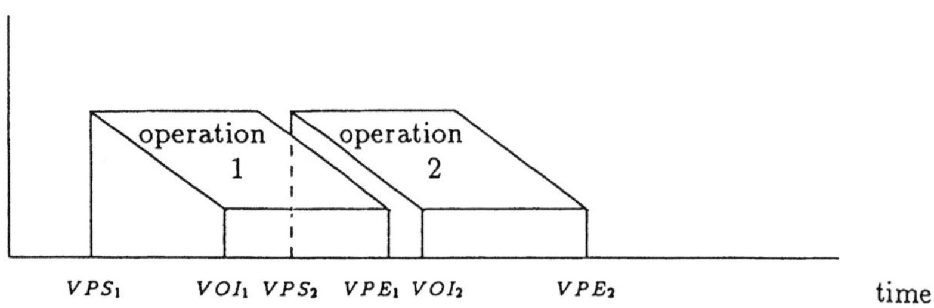

Fig.4-2 Start-up Time for Overlapped Operations

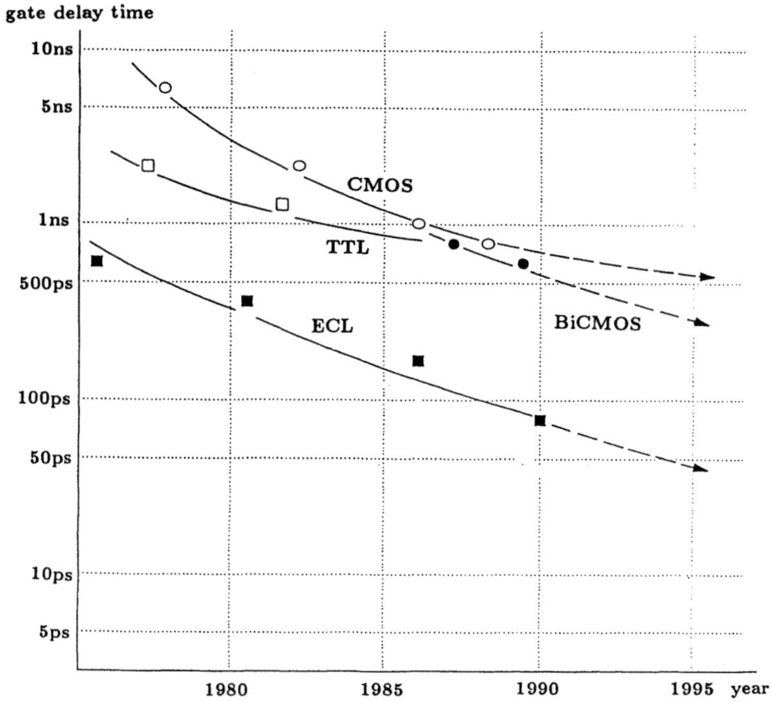

Fig. 5 Transition of the Speed of Logic Chip

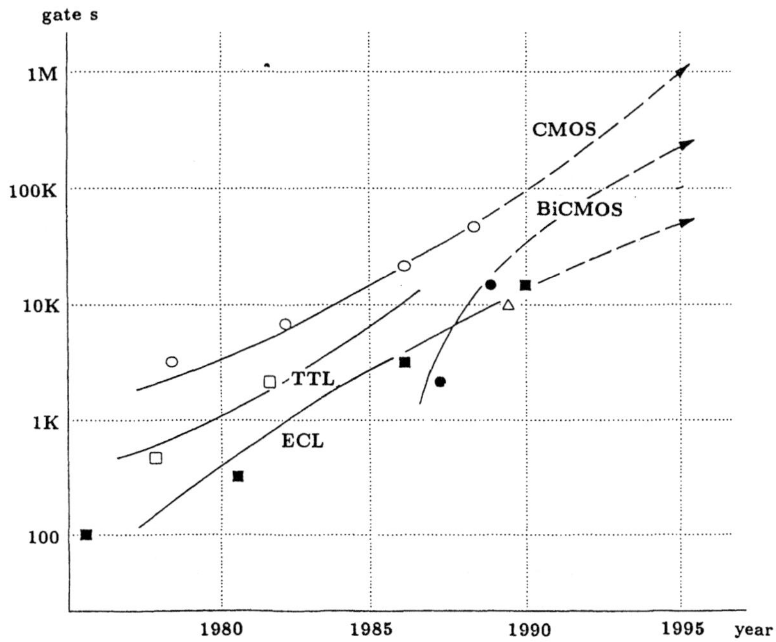

Fig. 6 Transition of the Density of Logic Chip

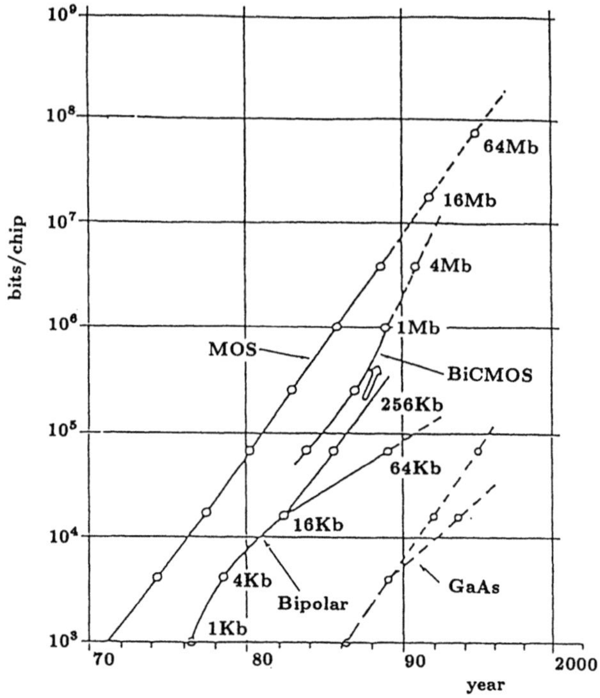

Fig. 7 Progress of Memory Chip

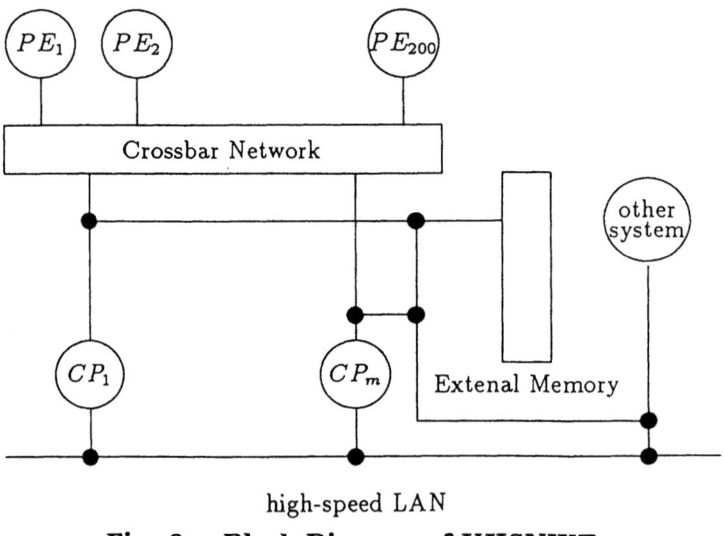

Fig. 8 Block Diagram of UHSNWT

IX ADDRESSES OF CONTRIBUTORS

Toshihiro Hongo

Cray Research Japan, Co., Ltd.
Ichibancho Eight-One Bldg., 3rd Floor
6-4, Ichibancho, Chiyoda-ku, Tokyo, 102 JAPAN

Keiichiro Uchida

Supercomputer Development Department, Main Frame Division,
Computer Systems Group, Fujitsu Limited
1015, Kamikodanaka Nakahara-ku, Kawasaki, 211 JAPAN

Shun Kawabe

Kanagawa Works, Hitachi, Ltd.
1 Horiyamashita, Hadano, Kanagawa, 259-13 JAPAN

Tadashi Watanabe

NEC Corporation
5-7-1, Shiba, Minato-ku, Tokyo, 108-01 JAPAN

Kenneth W. Neves

Boeing Computer Services
P.O. Box 24346, Seattle, WA 98034, U.S.A.

Kozo Fujii

Associate Professor
The Institute of Space and Astronautical Science
3-1-1, Yoshinodai, Sagamihara, Kanagawa, 229 JAPAN

Hideo Yoshihara

Former Liaison Scientist, Office of Naval Research
Roppongi, Tokyo, 106 JAPAN
Current address: 8620, NE, 20th St. Bellevue, WA 98004, U.S.A.

Wolfgang Gentzsch

College of Regensburg, Germany
Roentgenstr. 13, D-8402 Neutraubling, GERMANY

Masahiro Fukuda, Toshiyuki Iwamiya and Hajime Miyoshi

National Aerospace Laboratory
7-44-1, Jindaiji-Higashimachi, Chofu, Tokyo, 182 JAPAN

Addresses of the Editors of the Series "Notes on Numerical Fluid Mechanics"

Brief Instruction for Authors

Manuscripts should have well over 100 pages. As they will be reproduced photomechanically they should be typed with utmost care on special stationary which will be supplied on request.
In print, the size will be reduced linearly to approximately 75 per cent. Figures and diagrams should be lettered accordingly so as to produce letters not smaller than 2 mm in print. The same is valid for handwritten formulae. Manuscripts (in English) or proposals should be sent to the general editor, Prof. Dr. E. H. Hirschel, Herzog-Heinrich-Weg 6, D-8011 Zorneding.

GPSR Compliance

The European Union's (EU) General Product Safety Regulation (GPSR) is a set of rules that requires consumer products to be safe and our obligations to ensure this.

If you have any concerns about our products, you can contact us on ProductSafety@springernature.com

In case Publisher is established outside the EU, the EU authorized representative is:

Springer Nature Customer Service Center GmbH
Europaplatz 3
69115 Heidelberg, Germany

Batch number: 10577876

Printed by Printforce, the Netherlands